APPLIED PHYSICAL TECHNIQUES

APPLIED PHYSICAL TECHNIQUES

R. C. STANLEY, B.Sc., M.Inst.P.
Lecturer, Department of Applied Physics,
Brighton Polytechnic

CRANE, RUSSAK & COMPANY, INC.
NEW YORK

English edition first published in 1973 by
Butterworth & Co (Publishers) Ltd
88 Kingsway, London, WC2B 6AB

Published in the United States by
Crane, Russak & Company, Inc.
52 Vanderbilt Avenue,
New York, N.Y. 10017

Library of Congress Catalog Card No 72–96984

ISBN 0–8448–0169–0

Printed in Great Britain

PREFACE

There are many techniques in current use that in the past have been purely in the hands of the physicist or are based mainly on physical principles. They are techniques that were developed to deal primarily with research problems but, with the spread of technology into all fields of industry, these problems have become of more general interest with the techniques now becoming the everyday tools of the practising scientist and engineer. For example, the application of vacuum physics has now become of great importance to the electrical engineer in the production of thin film resistors and capacitors, and of micro-circuits. The techniques of optical microscopy are then used to examine the product and optical interference methods to measure the thickness of the films. Again in the electronic engineering industry, quartz crystals for oscillators must be orientated prior to slicing. This is done using X-ray techniques, which now come into a number of fields of engineering, for example, the mechanical engineer checking castings and welds for flaws and the civil engineer determining the nature of a corrosion product formed in the reactor of a nuclear power station. Similarly, electron microscopy, interference topography, and other techniques once used only in the realm of physics have now passed into common use.

This book, which may be considered as a collection of monographs on the more common techniques, is written primarily for undergraduate scientists and engineers who may well be called upon one day to use these techniques or at least interpret the results of others.

The physical foundations of each technique are presented, with emphasis being given to the practical aspects. Thus full descriptions are given of the practical applications and the types of apparatus and equipment involved, the types of problem or specimens to which each is most suited, and the form and assessment of the results that can be expected. The book is not meant to give an exhaustive study of each technique but rather to give the reader a good general working knowledge of a method so that he may more readily judge its potential to his own problem and assess the results of other workers.

Brighton R.C.S.

CONTENTS

One

VACUUM PHYSICS

1.1 Introduction

The applications of low-pressure physics have now become so diverse that a vacuum system can no longer be regarded as being solely a tool of the scientist, since applications may now be found in every branch of industry – from television tubes to the freeze drying of food. The prime importance of low pressures arises from the simple fact that the mean free path length, that is, the average distance travelled by a gas molecule, or for example a metal vapour atom, before a collison occurs with another gas molecule, can be increased by many orders of magnitude. Thus an electron's trajectory in a television tube or an electron microscope can be determined by an applied electric or magnetic field rather than by a scattering process due to random collisions with gas molecules.

The increase in the mean free path length has a wider application than merely to electron trajectories. If, for example, a metal is heated in a vacuum system to its molten state, or in some cases to just below its melting temperature, it evaporates. Atoms of the metal come off as a vapour and owing to thermal energies travel away from the hot metal source. Each of these atoms will travel a rectilinear path until such time as a collision occurs, whether it is with a gas molecule or with some solid object. With a typical vacuum system and the pressures involved in evaporation processes, the mean free path length for a metal vapour atom will be much greater than the dimensions of the vacuum system. Although some collisions must occur with gas molecules owing to the randomness of their motions, such collisions may be ignored in comparison with the vast number of atoms whose first collision is with a solid object. This solid object may well be the chamber walls but can also be any other object introduced into the chamber. Providing the object is not too hot, which would cause the metal atoms to be re-evaporated, the object will be coated with evaporated metal atoms which attach themselves one by one. In other words, any object or surface that 'sees' the hot metal source will be coated with that metal. It must be emphasised that any object can be coated in this way. For example, plastic car trim and plastic jewellery can be 'metallised', and glass surfaces can be coated with aluminium to form front reflecting mirrors for telescopes or for other optical processes. Not only metals can be evaporated but also a vast range of other substances as well. Thus quartz is evaporated onto front reflecting aluminised mirrors to protect the aluminium layer, and magnesium fluoride is evaporated onto lens surfaces to produce a non-reflecting inteference layer, termed blooming.

It must also be remembered that a long mean free path length means that a relatively long time will occur between collisions. This has relevance, for example, in oxidation studies since a sufficiently low vacuum enables a clean uncontaminated surface to be maintained for a convenient period. Thus oxidation processes owing to the arrival of oxygen molecules at the surface may be slowed down from virtually instantaneous changes occurring at atmospheric pressure to the same changes taking several hours under ultrahigh vacuum conditions.

Before, however, proceeding to discuss the production of a vacuum and its applications in greater detail, it is necessary to say a little concerning units and ranges of pressure. It has been customary to use a number of different units for the measurement of pressure. The first to achieve widespread usage was in terms of millimetres of mercury (mmHg), defined as the pressure exerted by a millimetre high column of mercury of density 13 595·1 kg m^{-3} in a place where the intensity of gravity is 9·80665 m s^{-2}. That is, 1 mmHg is equivalent to a pressure of $13{\cdot}5951 \times 9{\cdot}80665 = 133{\cdot}322$ N m^{-2}. This unit was superseded by the torr, defined as 1/760 of the standard atmospheric pressure. That is, 1 torr is equivalent to $101325/760 = 133{\cdot}322$ N m^{-2}. Thus for all practical purposes the mmHg and the torr are identical units; in fact they differ by less than 2×10^{-7} torr. Also used in this connection was the micron, equal to 10^{-3} torr.

Great Britain, in common with most of Europe, has now adopted SI units and consequently pressures are now measured in units of newtons per square metre (N m^{-2}), where 1 N m^{-2} is equivalent to 7·50064 $\times 10^{-3}$ torr (or mmHg). However, owing to the many existing pressure gauges with calibrations still in the older units, it may be expected that torr units will continue in common use for some considerable time to come. Consequently in this chapter torr equivalents will be given to the SI unit.

As to the total range of vacuum pressures, it has been found convenient to divide this into four approximate ranges, owing to the fact that different techniques are required in each case. These ranges are

Coarse vacuum: $10^5 - 10^3$ N m^{-2} (approx. $10^3 - 10^1$ torr)

Medium high vacuum: $10^3 - 10^{-1}$ N m^{-2} (approx. $10^1 - 10^{-3}$ torr)

High vacuum: $10^{-1} - 10^{-5}$ N m^{-2} (approx. $10^{-3} - 10^{-7}$ torr)

Ultrahigh vacuum: $10^{-5} - 10^{-9}$ N m^{-2} (approx. $10^{-7} - 10^{-11}$ torr)

KINETIC THEORY AND VACUUM

1.2 Kinetic Theory

The kinetic theory of gases is of especial interest in vacuum physics as it enables some useful relevant calculations to be made and at the same time gives a clearer insight into the problems involved in attaining a vacuum. By assuming a perfect gas in which infinitely small gas molecules of diameter d are moving in random directions due to thermal energy, it may be shown that the average distance travelled by a molecule of gas between successive collisions, termed the mean free path length l, is given by

$$l = \frac{1}{n\pi d^2 \sqrt{2}}$$

$$= \frac{1}{n\sigma\sqrt{2}} \tag{1.1}$$

where n is the number of molecules, each of diameter d, per unit volume, and $\sigma = \pi d^2$ is termed the collision cross-section. In practice, of course, the mean free path length is a very difficult quantity to measure since the diameter of a molecule has no exact physical meaning. It is nevertheless of importance that a mean free path length does exist even if it is not readily measurable.

Likewise from the kinetic theory of gases, it may be shown that the pressure p is given by

$$p = \tfrac{1}{3} mn\overline{c^2} = \tfrac{1}{3} \rho \overline{c^2} \qquad (1.2)$$

where m is the mass of a gas molecule and n is again the number of molecules per unit volume, so that $\rho = mn$ is the density of the gas at the pressure p. Here $\overline{c^2}$ is termed the mean squared velocity of the molecules, where c is the velocity of an individual molecule. Equation 1.2 may also be written in the form

$$pV = \tfrac{1}{3} Nm\overline{c^2} \qquad (1.3)$$

where $N = nV$ is the number of molecules in a volume V.

A further equation of interest is the ideal gas equation, which has been determined empirically as

$$pV = RT \qquad (1.4)$$

where p is the pressure in newtons per square metre, V is now the volume occupied by 1 kg-mole of gas, that is, by a mass of gas equal to the molecular weight expressed as an equivalent number of kilograms, R is the gas constant (equal to $8{\cdot}314 \times 10^3$ J kg-mole^{-1} K^{-1}), and T is the absolute temperature of the gas in kelvins. In pumping down a chamber to create a vacuum, it must be appreciated that a vast number of molecules must be removed and even when an ultrahigh vacuum has been obtained there still remains a vast number. To calculate how vast, recourse may be made to equation 1.4 which may be conveniently written for a volume in cubic metres as

$$pV = \frac{N}{N_A} RT \qquad (1.5)$$

where p is still the pressure measured in newtons per square metre, N is the total number of molecules in the volume V, now measured in cubic metres, and R and T are the gas constant and absolute temperature as before. N_A is Avogadro's number, the same for all substances, and is equal to $6{\cdot}023 \times 10^{26}$ molecules kg-mole^{-1}. Hence the number of molecules n in one cubic metre of gas is

$$n = \frac{N_A p}{RT}$$

$$= \frac{6{\cdot}023 \times 10^{26}}{8{\cdot}314 \times 10^3} \frac{p}{T}$$

$$= 7{\cdot}24 \times 10^{22} \frac{p}{T} \qquad (1{\cdot}6)$$

Thus, for example, at standard temperature and pressure, that is, at 0°C (273 K) and 101325 N m^{-2} (1 atm), the number of molecules per

Table 1.1 DIAMETERS OF SOME GAS MOLECULES, DETERMINED FROM VISCOSITY MEASUREMENTS

Gas	*Diameter* (*unit* = 10^{-10} m)
Argon	3·4
Carbon dioxide	3·9
Helium	2·6
Hydrogen	3·0
Nitrogen	3·7
Oxygen	3·5

cubic metre is $7{\cdot}24 \times 10^{22} \times 101325/273 = 2{\cdot}69 \times 10^{25}$. Even with an ultrahigh vacuum pressure of 10^{-9} N m^{-2} (approx. 10^{-11} torr), there are still some 10^{11} molecules per cubic metre. For pressures expressed in torr, n is equal to $9{\cdot}65 \times 10^{24}\ p/T$ molecules per cubic metre. How this number n varies at room temperature with the vacuum pressure is shown graphically in *Figure 1.1*. By assuming a diameter, for example, of $3{\cdot}7 \times 10^{-10}$ m for a molecule of nitrogen (*Table 1.1*) and substituting the value of n derived from equation 1.6 into equation 1.1, the mean free path length at an absolute temperature T and a pressure p is

$$l = \frac{2{\cdot}27 \times 10^{-5}\ T}{p} \text{ m} \qquad (1.7)$$

for a pressure p expressed in newtons per square metre, or

$$l = \frac{1{\cdot}70 \times 10^{-7}\ T}{p} \text{ m}$$

for a pressure expressed in torr. Assuming a value for room temperature of, say, 20°C (293 K), these expressions become in turn

$$l = \frac{6{\cdot}65 \times 10^{-3}}{p} \text{ m} \tag{1.8}$$

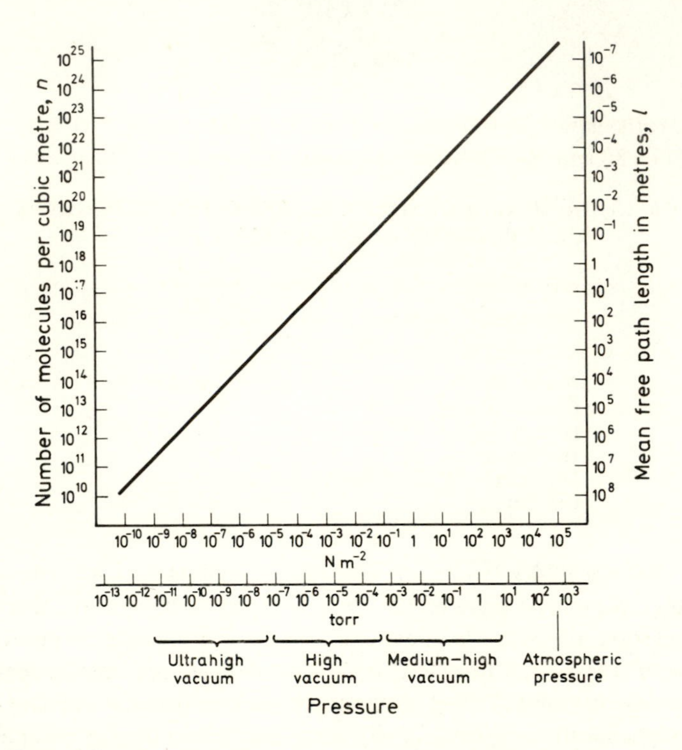

Figure 1.1. Variation of number of molecules and mean free path length of nitrogen at 20°C *with pressure*

where the pressure p is in newtons per square metre, and

$$l = \frac{4{\cdot}99 \times 10^{-5}}{p} \text{ m}$$

for a pressure expressed in torr. The variation of the mean free path length with pressure for nitrogen (and hence approximately air) at room temperature is shown graphically also in *Figure 1.1.*

It must also be appreciated that these gas molecules are travelling at great speeds. This can be derived by comparing equations 1.3 and 1.5, from which

$$\tfrac{1}{3} Nm\overline{c^2} = \frac{N}{N_A} RT$$

i.e.

$$\overline{c^2} = \frac{3RT}{N_A m} \tag{1.9}$$

where the symbols have their previous meanings. Then the 'root-mean-square' speed of the molecules is

$$c_{\text{r.m.s.}} = \sqrt{\overline{c^2}} = \sqrt{\frac{3RT}{N_A m}} \tag{1.10}$$

Considering, say, nitrogen, with a molecular mass of 2 x 14 atomic mass units, where one atomic mass unit is equal to $1{\cdot}660 \times 10^{-27}$ kg, and therefore with a molecular mass of $2 \times 14 \times 1{\cdot}660 \times 10^{-27} = 4{\cdot}65 \times 10^{-26}$ kg, the root-mean-square speed of these molecules at, say, 0°C (273 K) is

$$c_{\text{r.m.s.}} = \sqrt{\left(\frac{3 \times 8{\cdot}314 \times 10^3 \times 273}{6{\cdot}023 \times 10^{26} \times 4{\cdot}65 \times 10^{-26}}\right)}$$

$$= 4{\cdot}93 \times 10^2 \text{ m s}^{-1}$$

which is over 1100 miles per hour!

Another calculation of interest, that again gives an idea of the numbers involved, is for the time required to produce a monomolecular layer over unit area. This is of especial interest in oxidation studies. Knudsen has shown from kinetic theory that the number of gas molecules striking unit area of the chamber wall in unit time is given by

$$n' = \tfrac{1}{4} n\bar{c} \tag{1.11}$$

where n is the number of gas molecules in unit volume as before. It is assumed that the pressure is low so that the mean free path length of the gas molecules is great compared with the chamber dimensions. Here $\bar{c}$ is the arithmetic mean of the speeds of the molecules; it is not quite the same as $c_{\text{r.m.s.}}$ that has occurred in previous equations, such as equation 1.2. Thus $\bar{c}^2$ is not the same as $\overline{c^2}$. However, for the molecular speeds occurring in a gas these are not so very different, so that $c_{\text{r.m.s.}}$ is approximately equal to $1{\cdot}09\bar{c}$. That is, the root-mean-square speed of a gas molecule is about 9% greater than the average speed. Hence, substituting in equation 1.11 from equations 1.6 and 1.10

$$n' = \frac{1}{4}\left(\frac{N_\mathrm{A}p}{RT}\right)\frac{1}{1{\cdot}09}\sqrt{\frac{3RT}{N_\mathrm{A}m}}$$

$$= \frac{p}{4{\cdot}36}\sqrt{\frac{3N_\mathrm{A}}{RTm}} \tag{1.12}$$

Writing $m = M/N_\mathrm{A}$ where M is the molecular weight in atomic mass units, whereas m is the molecular weight in kilograms, equation 1.12 becomes

$$n' = \frac{N_\mathrm{A}p}{4{\cdot}36}\sqrt{\frac{3}{RTM}} \tag{1.13}$$

Thus, for example, the number of impacts per square metre per second occurring at a pressure of p newtons per square metre at room temperature, say, 20°C (293 K), and with nitrogen for which $M = 28$, is

$$n' = \left[\frac{6{\cdot}023 \times 10^{26}}{4{\cdot}36}\sqrt{\left(\frac{3}{8{\cdot}314 \times 10^3 \times 293 \times 28}\right)}\right] p$$

$$= 2{\cdot}9 \times 10^{22}\ p \text{ impacts m}^{-2}\ \text{s}^{-1} \tag{1.14}$$

For pressures measured in torr, n' becomes equal to $3{\cdot}87 \times 10^{24}\ p$ impacts per square metre per second.

Returning now to the original problem of determining the time for a monomolecular layer to be formed on a surface at room temperature, it may be assumed that, if a gas molecule is to attach itself to, say, a metal surface, it will do so at an existing atomic site. In other words, the gas layer forms a lattice-like structure that is an extension of the crystalline lattice of the metal. For a typical metal the interatomic distance is of the order $2{\cdot}5 \times 10^{-10}$ m, giving a possible $1{\cdot}6 \times 10^{19}$ atomic sites per square metre. If every gas atom that hits this surface attaches itself to an atom site, then the time t taken to fill every site, that is, to form a monomolecular layer one square metre in area, would be

$$t = \frac{1{\cdot}6 \times 10^{19}}{2{\cdot}9 \times 10^{22}p}$$

$$= \frac{5{\cdot}5 \times 10^{-4}}{p}\ \text{s} \tag{1.15}$$

where the pressure p is measured in newtons per square metre. For a pressure measured in torr, the time t would become $(4{\cdot}12 \times 10^{-6})/p$ s. For a high vacuum of, say, 10^{-4} N m^{-2} ($7{\cdot}5 \times 10^{-7}$ torr) this time will be about $5\frac{1}{2}$ s increasing to over 15 h at an ultrahigh vacuum pressure of 10^{-8} N m^{-2} ($7{\cdot}5 \times 10^{-11}$ torr). This then is an important reason for the use of ultrahigh vacuum equipment – it enables surfaces to remain clean and uncontaminated for a sufficient time to enable experiments to be carried out.

Not all molecules or surfaces, however, allow every incident molecule to remain attached. Some do not attach themselves in the first place and others detach themselves after a short time. For metals, what is termed a sticking coefficient is a measure of the probability of a gas molecule striking the surface and remaining for an infinite time. It has values mainly between 0·1 and 1.

1.3 Conductance and Pumping Speed

The initial rush of air through the connecting pipes as a vacuum pump is switched on to evacuate a chamber creates a turbulent flow condition. This turbulent flow, however, very soon ceases and streamline conditions prevail. In the realm of vacuum physics, this streamline condition is termed *viscous flow*. Then the mean free path length is still short relative to the dimensions of the tube diameter. Thus, from equation 1.8 or from *Figure 1.1*, it is seen that, at a pressure of about 0·3 N m^{-2} ($2{\cdot}25 \times 10^{-3}$ torr), the mean free path length of the gas molecules is about 0·02 m, which is probably a similar dimension to the diameter of the connecting pipes.

At lower pressures still, the mean free path length becomes great compared with the dimensions of the apparatus and the molecules move almost completely independently of one another. Collisions with the walls now become much more frequent than collisions with other molecules. Gas flow in this condition is termed *molecular flow*. At intermediate pressures, when the mean free path length is of the same order as the diameter of the pipe through which the gas flows, the flow is dependent on both molecular and viscous properties and the flow is then said to be in the transition range.

In designing a vacuum system, it is important that the dimensions of the connecting pipes should be compatible with the type of flow and the pressure range involved, otherwise unduly long pump-down times may result. The capacities of the pumps should similarly be consistent with the dimensions of the system and the volume of the chamber to be evacuated. The main criterion in this context is the *conductance, C,* which may be expressed as

$$C = \frac{\text{pressure at any point} \times \text{volume flow at that point}}{\text{pressure drop}} \quad (1.16)$$

where the flow is expressed in cubic metres per second measured at the particular pressure at that point. Hence what may be termed the 'throughput', Q, can be expressed as

$$\begin{aligned} Q &= C(p_1 - p_2) \\ &= \text{pressure} \times \text{volume flow} \end{aligned} \quad (1.17)$$

where $(p_1 - p_2)$ is the pressure drop over which the conductance is measured. The units of throughput will be newton metre per second ($N\ m\ s^{-1}$). For viscous flow, when the mean free path length is small relative to the diameter of the conducting pipe, the flow through the pipe is in accordance with the Hagen–Poiseuille equation for gases. Thus the conductance of a pipe of length l, diameter d, and with a throughput Q, equal to the product of the volume flow and pressure, is given by

$$\begin{aligned} C &= \frac{Q}{p_1 - p_2} = \frac{\pi (p_1 + p_2) d^4}{256 \eta l} \\ &= \frac{\pi p d^4}{128 \eta l} \end{aligned} \quad (1.18)$$

where η is the viscosity of the gas at the appropriate temperature and p is the mean pressure along the pipe, that is, $p = (p_1 + p_2)/2$ where p_1 and p_2 are the pressures at the two ends. Assuming a viscosity of air at 20°C (293 K) of $\eta = 1{\cdot}81 \times 10^{-5}\ N\ s\ m^{-2}$, equation 1.18 simplifies to

$$C = 1{\cdot}35 \times 10^3 \frac{pd^4}{l}\ m^3\ s^{-1} \quad (1.19)$$

for the conductance of a pipe at a pressure p measured in newtons per square metre. This expression may be written as

$$C = 1{\cdot}80 \times 10^5 \frac{pd^4}{l}\ m^3\ s^{-1}$$

if the pressure is measured in torr.

For molecular flow, when the mean free path length is great compared to the diameter of the pipe, Knudsen has derived on a partly empirical basis values for conductances. These are given in *Table 1.2*

Table 1.2 FORMULAE FOR CALCULATING CONDUCTANCES, IN UNITS OF $m^3\ s^{-1}$

Viscous conductance	
round pipe	$1{\cdot}35 \times 10^3\ pd^4/l$
Molecular conductance	
round pipe	$121\ d^3/(l + 1{\cdot}3d)$
orifice – any shape	$116\ A$

Pressure, p,	unit = $N\ m^{-2}$
Diameter of pipe, d,	unit = m
Length of pipe, l,	unit = m
Orifice area, A,	unit = m^2

for some common situations for gas temperatures of 20°C (293 K). It will be seen that molecular conductances are independent of the pressure. For high conductance it is seen from the table that it is essential to employ a connecting pipe as short as possible and with as large a diameter as possible. If the conductance is limited by the pipe, then no matter how great the capacity of the pump, virtually no faster pumping can be achieved.

Conductance formulae are not necessary for the transition range between mixed molecular and viscous flow. It is found to be quite sufficient to use the formulae for viscous flow conductance down to a transition pressure p_t, where

$$p_t = \frac{7{\cdot}5 \times 10^{-2}}{d}\ N\ m^{-2} \tag{1.20}$$

$$= \frac{5{\cdot}5 \times 10^{-4}}{d}\ \text{torr}$$

and where d is the diameter of the pipe in metres, and then to use molecular flow conductances for lower pressures. The transition range in fact extends from $10p_t$ to $0{\cdot}1p_t$. For systems in which different flow conditions prevail, it is found that

$$\text{total conductance} = \text{molecular conductance} + \text{viscous conductance} \tag{1.21}$$

Also, if several conductors are in series, the total conductance C is given by

$$\frac{1}{C} = \frac{1}{C_1} + \frac{1}{C_2} + \frac{1}{C_3} + \ldots \tag{1.22}$$

where $C_1, C_2, C_3, \ldots$, are the individual conductances. Similarly, for conductors in parallel

$$C = C_1 + C_2 + C_3 + \ldots \tag{1.23}$$

A pumping speed S_p of a pump can also be defined, by the expression

$$Q = p\,S_p \tag{1.24}$$

where the throughput Q, by comparison with equation 1.17, is equal to the product of the pressure p, in this case measured at the input to the pump, and the volume per unit time of gas being removed by the pump, again measured at the pressure of the pump inlet. In other words, the pumping speed is the volume of gas removed by the pump in unit time, measured at the pressure of the pump inlet. If the pump is separated from the chamber by a pipe or orifice of conductance C, the effective speed of the pump becomes S, where

$$\frac{1}{S} = \frac{1}{S_p} + \frac{1}{C} \tag{1.25}$$

As an example the problem of evacuating a small electronic valve may be considered. Owing to its size only a small bore pipe connection to the pump is possible. From equation 1.25, the effective pumping speed at the chamber is given by

$$S = \frac{S_p C}{S_p + C} \tag{1.26}$$

where S_p is the pumping speed of the pump itself and C is the conductance of the connecting pipe. Supposing a pump is chosen which has a speed 10 times the conductance of the pipe, that is, $S_p = 10C$, then from equation 1.26 the effective speed is

$$S = \frac{10}{11}C$$

If instead a pump with 10 times the speed of the first one had been used, so that $S_p = 100C$, the effective speed would be increased to

$$S = \frac{100}{101} C$$

This is only an increase of about 9% in the effective speed for a ten-fold increase in the pump speed, so that the extra expense of the larger pump would not be justified.

From Boyle's law, for a small change δp in pressure with its corresponding volume change δV,

$$pV = (p - \delta p)(V + \delta V) \tag{1.27}$$

Hence

$$\delta p = \frac{p}{V} \delta V$$

and

$$\frac{dp}{dt} = \frac{p}{V} \frac{dV}{dt} \tag{1.28}$$

However, by definition the speed S is given by

$$S = -\frac{dV}{dt} \tag{1.29}$$

where the negative sign arises since the volume removed decreases as time increases, that is, the speed gets less with time. Then, from equations 1.28 and 1.29, the rate of reduction of pressure is

$$\frac{dp}{dt} = -\frac{S}{V} p \tag{1.30}$$

At the ultimate vacuum pressure of the system, that is, when the pressure reaches an equilibrium state due to the limiting nature of the the pump and any leaks, dp/dt becomes zero and therefore so does the pumping speed S from equation 1.30.

Now, rewriting equation 1.30 as

$$S \, dt = -\frac{V}{p} dp \tag{1.31}$$

and integrating, gives

$$S \int_{t_1}^{t_2} dt = -V \int_{p_1}^{p_2} \frac{1}{p} dp \tag{1.32}$$

where p_1 and p_2 are pressures measured at times t_1 and t_2, and the pumping speed S is assumed constant over this pressure range. Therefore

$$S = \frac{V}{t_2 - t_1} \log_e \left(\frac{p_1}{p_2}\right)$$

$$= \frac{2{\cdot}303\, V}{t_2 - t_1} \log_{10} \left(\frac{p_1}{p_2}\right) \qquad (1.33)$$

This equation enables the speed S to be measured conveniently, in units of $m^3\ s^{-1}$, since the pressures and corresponding times are easily measured together with the volume being evacuated, $V\ m^3$. Conversely, knowing the speed over this pressure range, the time required to reduce the pressure from p_1 to p_2 can be calculated. It should be remembered that S is the effective speed of the pump at the chamber, taking into account the conductance of the pipe connections as in equation 1.26.

As an example of the use of the above equations in assessing the performance of a vacuum system, consider a vacuum chamber connected to a pump with a pumping speed of $5 \times 10^{-2}\ m^3\ s^{-1}$ by a pipe 0·02 m in diameter and 0·2 m long and with a lowest pressure of $1 \times 10^{-4}\ N\ m^{-2}$ attainable at the pump inlet. First the conductance of the pipe should be calculated. From equation 1.20, the transition pressure between viscous and molecular conductance is

$$p_t = \frac{7{\cdot}5 \times 10^{-2}}{0{\cdot}02}$$

$$= 3{\cdot}75\ N\ m^{-2}$$

Obviously the pressures in the system are well below this and the conductance is therefore (from *Table 1.2*)

$$C = \frac{121 d^3}{l + 1{\cdot}3d}$$

$$= \frac{121 \times 0{\cdot}02^3}{0{\cdot}2 + 1{\cdot}3 \times 0{\cdot}02}$$

$$= 4{\cdot}3 \times 10^{-3}\ m^3\ s^{-1}$$

From equation 1.26 the effective pumping speed at the chamber is

$$S = \frac{S_p C}{S_p + C}$$

$$= \frac{(5 \times 10^{-2})(4{\cdot}3 \times 10^{-3})}{(5 \times 10^{-2}) + (4{\cdot}3 \times 10^{-3})}$$

$$= 4{\cdot}0 \times 10^{-3}\ \mathrm{m^3\ s^{-1}}$$

Now, in agreement with Boyle's law,

$$pS = p_p S_p$$

where p is the pressure at a point where the speed, that is, the volume flowing per second, is S. Here p_p and S_p are the corresponding pressure and speed at the pump inlet. Then the pressure at the chamber is p where

$$p = \frac{p_p S_p}{S}$$

$$= \frac{(1 \times 10^{-4})(5 \times 10^{-2})}{4{\cdot}0 \times 10^{-3}}$$

$$= 1{\cdot}3 \times 10^{-3}\ \mathrm{N\ m^{-2}}$$

Alternatively, using equation 1.16

$$\text{pressure drop} = \frac{\text{pressure at any point x volume flow at that point}}{\text{conductance, } C}$$

$$= \frac{(1 \times 10^{-4})(5 \times 10^{-2})}{4{\cdot}3 \times 10^{-3}}$$

$$= 1{\cdot}2 \times 10^{-3}$$

$$= 12 \times 10^{-4}\ \mathrm{N\ m^{-2}}$$

This is the pressure drop along the pipe and hence the pressure in the chamber is

$$p = (1 \times 10^{-4}) + (12 \times 10^{-4})$$

$$= 13 \times 10^{-4}$$

$$= 1{\cdot}3 \times 10^{-3}\ \mathrm{N\ m^{-2}}$$

as before.

The time involved in pumping down the chamber can also be calculated. Supposing, for example, it is required to reduce the pressure in the chamber from 10^{-1} N m^{-2} to 10^{-3} N m^{-2}, assuming the pumping speed remains substantially constant over this pressure range, where the chamber has a volume of, say, 0·1 m^3. Then from equation 1.33 the time taken will be

$$t_2 - t_1 = \frac{2{\cdot}3V}{S} \log_{10}\left(\frac{p_1}{p_2}\right)$$

$$= \frac{2{\cdot}3 \times 0{\cdot}1}{4{\cdot}0 \times 10^{-3}} \log_{10}\left(\frac{10^{-1}}{10^{-3}}\right)$$

$$= 1{\cdot}2 \times 10^2 \text{ s}$$

$$= 2 \text{ min}$$

VACUUM SYSTEMS

1.4 Vacuum Gauges

Before describing actual pumps and complete pumping systems, it is perhaps more fitting to deal first with gauges that are used to measure the pressures produced in vacuum systems. Once a pressure is reached below that which may be practically measured by a manometer, there is a lack of absolute gauges. In fact the only gauge for vacuum pressures that can be regarded as completely absolute is the *McLeod gauge*. Its operation is based purely on Boyle's law and it is against this gauge that all other low-pressure gauges are calibrated, often with considerable extrapolation.

Figure 1.2 (a) shows an arrangement of the gauge. With it evacuated, a column of mercury some 760 mm high can be supported and consequently facilities must exist for lowering the mercury reservoir this amount to clear the mercury from the gauge. An alternative arrangement for doing this shown in *Figure 1.2 (b)* uses a vacuum pump to reduce this head of mercury. To measure a pressure the mercury column is raised by raising the mercury reservoir or by allowing air into the mercury flask in the alternative arrangement. When the mercury reaches the level aa, it traps in chamber A a volume V of gas at the vacuum pressure p to be measured. The mercury is allowed to rise to the height b in the comparison tube [*Figure 1.2 (c)*], which is level with the top of the closed capillary tube. The mercury in the closed

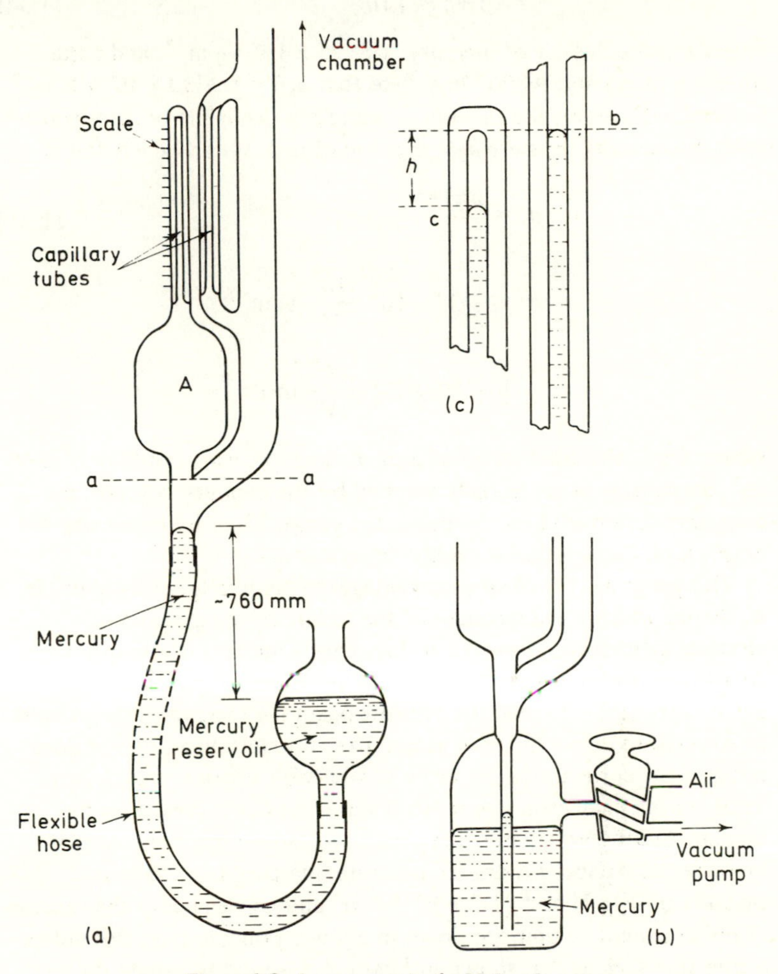

Figure 1.2. Arrangement of a McLeod gauge

capillary tube then reaches some level c, where the difference in levels c to b is h metres. Thus the volume V of gas at pressure p has now been compressed to some new volume $V' = Ah$, where A is the cross-sectional area of the capillary tube, with a pressure p' equal to the vacuum pressure p plus the pressure exerted by the mercury column of height h. Therefore, by Boyle's law,

$$\begin{aligned} pV &= p'V' \\ &= (p + \rho gh)Ah \end{aligned} \tag{1.34}$$

where ρ is the density of mercury, $1{\cdot}3595 \times 10^4$ kg m^{-3}, and g the intensity of gravity, $9{\cdot}8067$ m s^{-2}, so that $\rho gh = 1{\cdot}3332 \times 10^5 h$ N m^{-2}. Generally, however, the vacuum pressure p is negligible in comparison with the mercury pressure ρgh, and equation 1.34 can be written as

$$p = \frac{\rho g A h^2}{V} \tag{1.35}$$

$$= 1{\cdot}33 \times 10^5 \frac{Ah^2}{V} \text{ N m}^{-2}$$

$$= 9{\cdot}97 \times 10^2 \frac{Ah^2}{V} \text{ torr}$$

where A m^2, the cross-sectional area of the measuring capillary, and V m^3, the volume of gas initially trapped by the mercury column, are constants determined for the particular gauge. Thus, by measuring the height h m, the pressure is readily determined.

This gauge has the obvious advantages of simplicity, particularly in its theory which is independent of the nature of the gas, and of absolute calibration. However, it does have a number of disadvantages. Apart from the fact that it cannot read pressures continuously or instantaneously – to read the pressure again the mercury column must be lowered to below the level aa and then brought up again – it does not give the correct pressure when condensable vapours are present. Thus considerable error can result if water vapour is present in the system. At low enough pressures, mercury vapour itself becomes of importance. At about room temperature, mercury has a vapour pressure of $0{\cdot}16$ N m^{-2} ($1{\cdot}2 \times 10^{-3}$ torr) and consequently the vacuum chamber cannot be pumped down to a lower pressure than this unless a cold trap such as that shown in *Figure 1.3* is used to isolate the McLeod gauge from the rest of the vacuum system.

Owing to the pumping action of the cold trap, produced by the movement of mercury vapour from the gauge to the trap carrying along some molecules of gas, the pressure at the gauge becomes lower than at the trap. Thus at pressures lower than about 10^{-2} N m^{-2} (approx.

10^{-4} torr) errors begin to be introduced, although they can be reduced by inserting a short length of capillary tubing between the gauge and the trap. This reduces the conductance sufficiently to make negligible

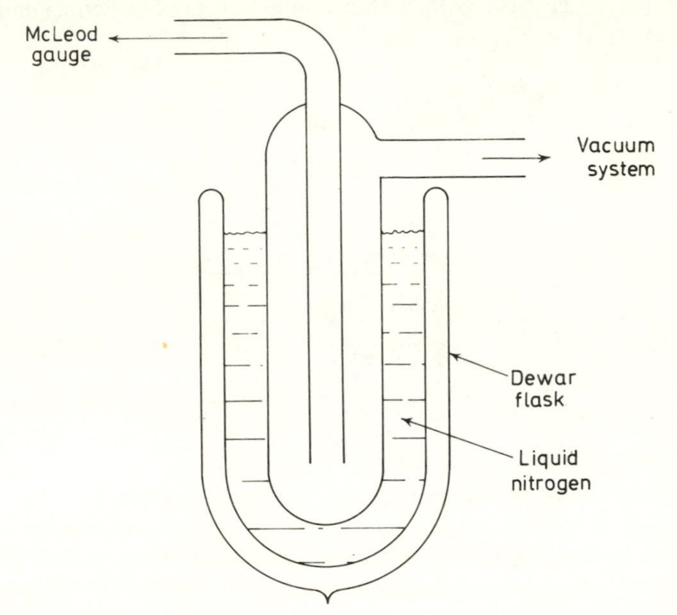

Figure 1.3. Design of a cold trap to prevent mercury vapour from a McLeod gauge entering the rest of the vacuum system

the pumping action due to the flow of mercury vapour, although at the same time it does increase the time taken for the pressure of the gas in the McLeod gauge to equalise to that of the rest of the system after a pressure measurement has been made. A more serious disadvantage is the lower limit of pressure measurable. This is set mainly by the volume trapped by the mercury in the gauge, since a large volume may become an appreciable part of the total volume of the system to be evacuated by the pump, as well as leading to a cumbersome gauge requiring a large volume of mercury with all its associated weight.

The sensitivity of the measuring capillary cannot be increased appreciably as a decrease in its cross-sectional area leads to 'sticking' and breaking of the mercury thread. Thus the McLeod gauge, although it is absolute, is not very satisfactory for routine use and is mainly used to calibrate other more convenient gauges. Its useful working range is shown in *Figure 1.4* in comparison with other commonly used gauges.

A very useful gauge is the *Pirani gauge*. This takes over from the mercury manometer, or as is more usual, it is used as the first vacuum gauge of the system since in practice the pressure is down to the Pirani range *(see Figure 1.4)* within a minute or so of switching on the

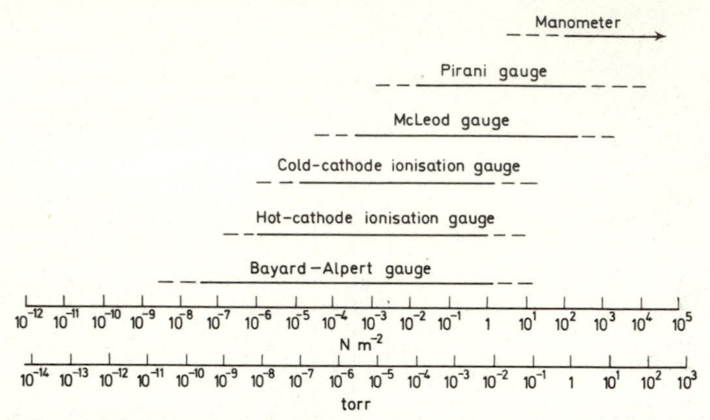

Figure 1.4. Ranges of pressures measurable with commonly used gauges – the full line indicates the usual range

pumps. The gauge element is simply a hot filament which is generally heated only to a dull red to prevent burn-out. Changes in gas pressure affect the rate at which heat is conducted away from the filament and consequently affect its temperature. The filament material is chosen to have a high temperature coefficient of resistance so that changes in the gas pressure manifest themselves as changes in resistance measurable by a Wheatstone bridge arrangement as in *Figure 1.5*.

So that ambient temperature changes shall not affect the results, an identical filament sealed off in a vacuum container is sometimes used as a compensator. By connecting this compensator to the adjacent arm in the Wheatstone bridge, ambient effects are cancelled. The usual method of using the bridge is not by null balancing but by applying a constant voltage of the order of 10 V across the bridge and adjusting the 'set zero' resistance *(Figure 1.5)* for balance when the hot filament is in a vacuum lower than its operating range. The galvanometer G then reads zero – the lowest-pressure reading. At higher pressures the balance is upset and the galvanometer G then records an out-of-balance current which is a measure of the gas pressure. A calibration is then required between the pressure as measured by a McLeod gauge and this current, to enable operation as a continuously reading gauge. It is usual to use one meter instead of the two shown in *Figure 1.5*, with a suitable switching arrangement so that the one meter is first used for setting the bridge voltage and then for the pressure readings.

The upper pressure limit is due to the thermal conductivity of the gas becoming mainly independent of the pressure at higher values, whereas the lower limit is due to radiation losses becoming relatively more important than conduction by the gas. Since the calibration of

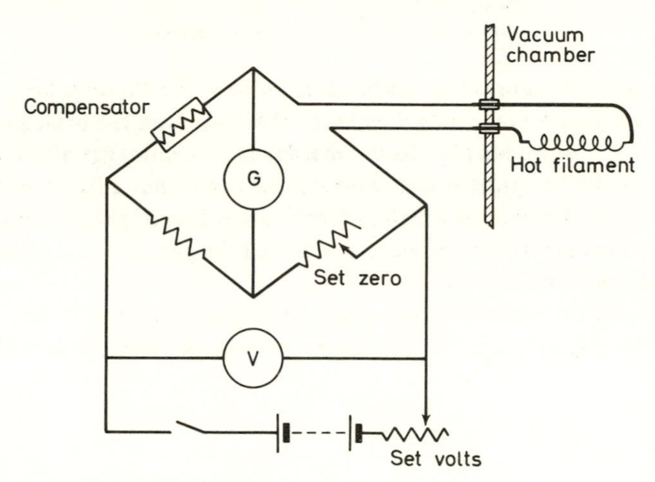

Figure 1.5. A Wheatstone bridge arrangement suitable for use with a Pirani gauge

the gauge is dependent on the thermal conductivity of the gas at a particular pressure, the calibration is upset when the gauge is used with a gas of different thermal conductivity. The true pressure must then be determined from the expression

$$\text{pressure} = \frac{\text{gauge reading}}{\text{calibration factor}} \tag{1.36}$$

where the calibration factor for some common gases is given in *Table 1.3*. The gauge is normally calibrated with varying pressures of nitrogen gas, so that the factors are recorded relative to this. It is seen from the table that butane is sufficiently different to cause a pronounced change in the gauge reading if present and is therefore useful as a 'search gas' in locating a leak in the vacuum system. The method is to play a small jet of the gas onto likely locations for leaks, such as pipe unions and taps, when a sudden change in the Pirani gauge reading will indicate that the butane jet has found the region of the leak.

Table 1.3 CALIBRATION FACTORS FOR SOME COMMON GASES USED WITH PIRANI GAUGES

Butane	2·5
Carbon dioxide	1·1
Carbon monoxide	1·0
Hydrogen	1·3
Mercury vapour	0·3
Neon	0·9
Nitrogen	1·0 – reference

For lower pressures than those measurable with a Pirani gauge, ionisation gauges come into their own. These work on the principle of producing positive ions by the bombardment of neutral gas molecules by electrons. The positive-ion current produced is thus a function of the number of collisions which in turn is a function of the pressure. Two sources are used to produce the bombarding electrons – a cold cathode or a hot filament.

The *cold-cathode* ionisation gauge, termed a *Penning* or *Philips* (after the original manufacturer) *gauge*, comprises a ring anode with a

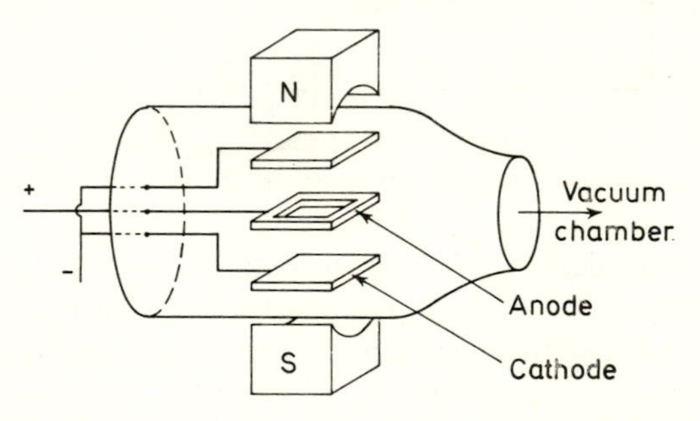

Figure 1.6. Arrangement of a Penning ionisation gauge

cathode mounted on either side as in *Figure 1.6*. The anode–cathode assembly is enclosed within a glass or a metal envelope connected to the vacuum chamber. A development of this basic arrangement is to use a flattened metal tube at earth potential as the cathode with an anode ring mounted inside the tube. The cathode metal is often aluminium, nickel-plated copper, or stainless steel. A d.c. voltage of about 2 kV applied between the anode and cathode causes any free electrons in the region to move towards the anode. These will collide with gas molecules and produce ionisation. The heavy positive ions produced will move towards the cathode causing further ionisation on their way and on collision with the cathode will liberate more electrons, which in turn travel towards the anode and produce more ion pairs on their journey. Thus there is a positive-ion current towards the cathode and a negative-ion (electron) current towards the anode. If the anode and cathode are of different surface areas, then conduction will take place more easily in one direction than in the other. In the Penning gauge the anode has a relatively small surface area, which means that there is a much greater current density on its surface. If this current density exceeds a certain critical value on the small anode, but not on the larger cathode, it would require a higher voltage between the anode and cathode to maintain a negative-ion current towards the anode than would be required for a positive-ion current of the same magnitude towards the cathode. Consequently there is a greater positive-ion current than there is a negative one. The magnitude of this positive-ion current is of course dependent on the number of ion pairs produced which in turn is dependent on the number of molecules of gas present, that is, on the pressure.

However, at low pressures the mean free path length is great and therefore steps must be taken to ensure that collisions do occur. This is done by fitting a magnet with a flux density of about 5×10^{-2} T to the gauge, as in *Figure 1.6*. Thus electrons moving towards the anode move in spiral paths and many pass right through the ring anode to be repelled again by the cathode. This may happen many times, the electrons spiralling back and forth through the anode before being finally captured, and hence travelling paths that may be several hundred times the direct path between a cathode and the anode. Again, as a gauge, it must be calibrated against a McLeod gauge and since its lower limit is below that of the McLeod some extrapolation is required. For pressures below about 10^{-6} N m^{-2} (approx. 10^{-8} torr), the relatively small number of gas molecules allows few collisions and the ion current becomes too small for practical measurement but, above this pressure, owing to the ruggedness of the construction and the lack of any hot filament to burn out, the gauge is found to be extremely useful and popular. There is, however, an upper limit to its pressure range of about 3 N m^{-2} (approx. 2×10^{-2} torr). This is due to the breakdown of carbon compounds, for example from pump oils, which causes contamination of the electrode surfaces, although the gauges are usually demountable and easily cleaned. To avoid contamination and the possibility of arcing, the gauge should not be left on when at atmospheric pressure.

There are some minor disadvantages with the Penning gauge. If switched on at atmospheric pressure there will normally be no discharge or ionisation and consequently there will be no current flow. The gauge will then erroneously indicate a very good vacuum. At lower pressures, within the operating range of the gauge, the existence of some free electrons is required to initiate the ionisation process. The existence of these free electrons is usually due to some external agency, such as cosmic rays, but if by chance they are absent some difficulty may be experienced in starting the discharge. The gauge also acts as a pump and must therefore have a very high conductance connection to the vacuum

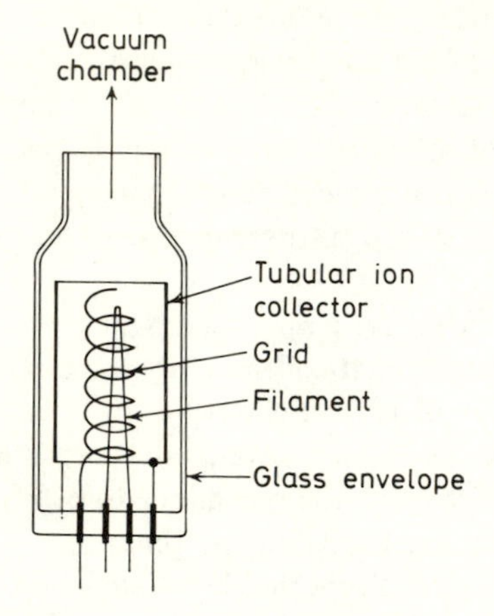

Figure 1.7. Section through a hot-cathode ionisation gauge

chamber whose pressure is required, otherwise an anomalously low pressure will be registered. This pumping action arises from the bombardment of the cathode by heavy positive ions, causing metal atoms to be knocked, or sputtered, from the cathode surface. These metal atoms are deposited over the gauge walls and by chemical reaction with the gas molecules effectively remove them from the free volume and hence reduce the pressure.

A gauge working on a similar principle but having the advantage of greater accuracy is the *hot-cathode ionisation gauge*. Its construction is that of a conventional triode valve, as in *Figure 1.7* where a central tungsten filament or emitter is surrounded by a grid of molybdenum wire. Around this is a nickel cylinder which acts as an ion collector. A positive voltage of about 150 V is applied to the grid, causing electrons emitted by the hot filament to be accelerated towards the grid and to produce ionisation by collision with the gas molecules in their passage.

Owing to the open structure of the grid, some of the electrons will pass through the grid to cause ionisation beyond, oscillating back and forth before final capture by the grid. The positive ions produced by ionisation of gas molecules go to the collector which carries a voltage of about 20 to 30 V negative with respect to the filament. The constancy of electron production by the filament is maintained by adjusting the filament temperature by varying the voltage, to maintain a constant emission current between the filament and the grid of about 1–5 mA.

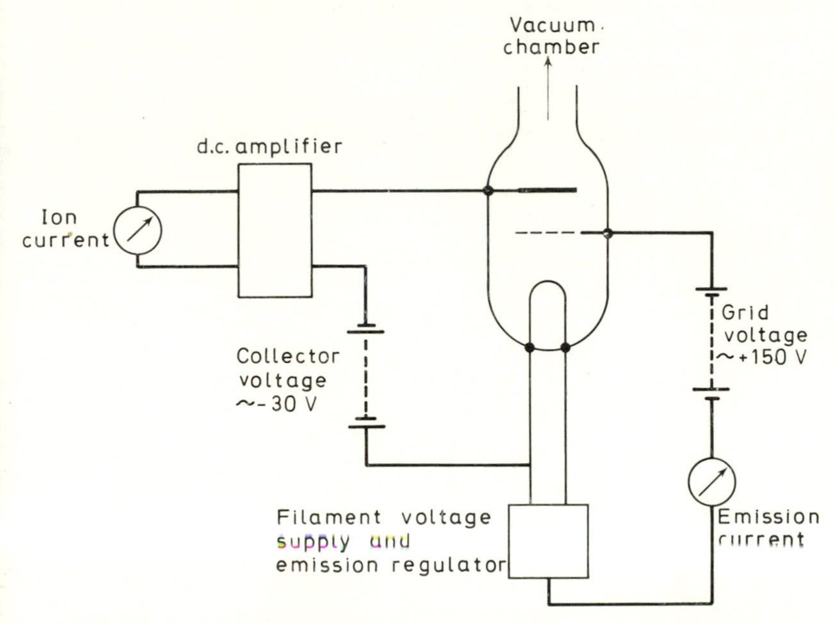

Figure 1.8. The electrical circuit for a hot-cathode ionisation gauge, shown diagrammatically

The pressure, which is a function of the number of positive ions produced, is determined from the ion current to the collector. It is usual for the emission current to be maintained constant electronically and the ion current to be passed to a d.c. amplifier before being registered by a meter, rather than simply using a microammeter, as in *Figure 1.8*.

An advantage over a cold-cathode ionisation gauge, apart from the greater accuracy, is the ease of outgassing. Molecules of gas tend to stick to cold surfaces and come off slowly with time; consequently they give the same effect as a small leak. By heating the surfaces these gas molecules can be driven off, or outgassed, and pumped away to cause no further trouble. To outgas a cold-cathode ionisation gauge a surrounding oven is required, whereas a hot-cathode ionisation gauge can be outgassed by passing an increased current through the filament to heat the gauge. Again, as with the cold-cathode ionisation gauge, a high-conductance connection is required to the chamber of which the gas pressure is required, as the gauge shows some pumping action due to gettering, that is, the using up of free gas molecules by chemical reaction with metal surfaces and atoms. For this reason a nude gauge, that is, one without an envelope, is often used and mounted inside the actual vacuum chamber.

The upper limit of the gauge is about 1 N m^{-2} (approx. 10^{-2} torr). In the region of this pressure physical deterioration of the filament is

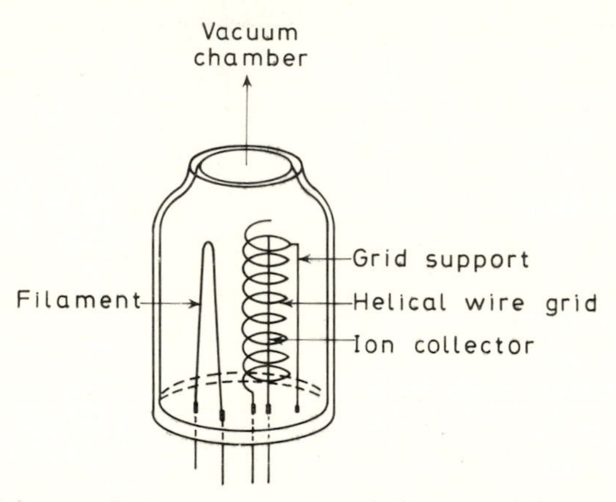

Figure 1.9. Bayard–Alpert arrangement of a hot-cathode ionisation gauge

rapid and it is soon likely to burn out. Thus the gauge must never be switched on when at atmospheric pressure. Owing to the usual electronic amplification of the ion current, the copious supply of electrons produced by the hot filament, and the open structure of the grid, a magnet to increase the path length of the electrons before capture is not generally required.

The lower limit of pressure is found to be about 10^{-6} N m^{-2} (approx. 10^{-8} torr), when the corresponding ion current is about 10^{-10} A. There is no great difficulty in measuring this order of current providing stray capacitance effects are avoided by, for example, using a well-screened coaxial cable connection between the ion collector and the amplifier. In practice, the lower pressure limit is set by the electrons striking the grid and causing it to emit soft X-rays. These X-rays in turn irradiate the ion collector causing it to emit secondary electrons so that the positive-ion current as recorded by the gauge appears greater by the amount of this electron current from the collector. At pressures in the region of 10^{-6} N m^{-2} (10^{-8} torr), this residual electron current is comparable to the pressure-related positive-ion current and at lower pressures becomes the overriding factor. Thus to extend the lower limit it is necessary to reduce this X-ray effect.

In the *Bayard–Alpert gauge* the configuration of the hot-cathode ionisation gauge is modified by reducing the ion collector in size from a cylinder to a wire as in *Figure 1.9*, so as to reduce the area susceptible to X-radiation and so reduce the secondary electron emission from the collector. The gauge envelope is of glass or stainless steel, with the metal version suitable for baking in an oven for outgassing, as is required for ultrahigh vacuum use, and both versions may also be outgassed by passing a greater current through the filament.

With this arrangement the lower limit of pressure measurement set by X-ray emission is reduced to about 3×10^{-8} N m^{-2} ($2{\cdot}2 \times 10^{-10}$ torr). The upper pressure limit of about 1 N m^{-2} (approx. 10^{-2} torr) remains unchanged with the new configuration and is still set mainly by the risk of filament burn-out. An extension of about another order lower on the lowest pressure measurable can be achieved by using another single wire electrode close to the grid and parallel to the ion collector. By varying the potential on this electrode between zero, with respect to the filament, and the potential of the grid, the positive-ion current to the ion collector is modulated and can be selectively discriminated against with respect to the constant residual current due to X-ray emission.

As with the Pirani gauge, an ionisation gauge has different sensitivities for different gases, this time owing to differences in their ionisation characteristics. If the gauge is calibrated with nitrogen against a McLeod gauge the correct pressure can be calculated from the expression

$$\text{pressure} = \frac{\text{gauge reading}}{\text{calibration factor}} \qquad (1.37)$$

where values of the calibration factors for various gases are given in *Table 1.4*. Thus hydrogen in the form of coal gas makes a cheap and convenient search gas, provided the usual safety precautions are taken, although helium would be better, apart from its cost, and safer.

Table 1.4 CALIBRATION FACTORS FOR USE WITH AN IONISATION GAUGE

Butane	10·0
Carbon dioxide	1·37
Carbon monoxide	1·05
Helium	0·14
Hydrogen	0·4
Mercury vapour	2·7
Neon	0·25
Nitrogen	1·0 – reference
Oxygen	0·8
Water vapour	2·0

For more accurate work at lower pressures, recourse must be made to *mass spectrometers*. A typical arrangement of a 180° mass spectrometer suitable for vacuum work is shown diagrammatically in *Figure 1.10*, where the whole spectrometer is integral with the vacuum chamber. It is generally made of stainless steel so that it can be baked

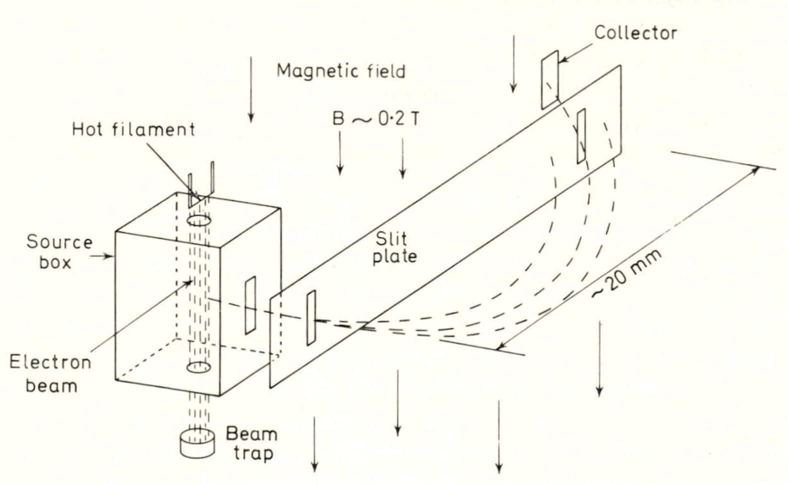

Figure 1.10. Arrangement of 180° *mass spectrometer for vacuum measurements*

together with the rest of the vacuum chamber for outgassing purposes, although it is usually necessary to remove the magnet first. A constant-emission hot filament is used to supply a beam of electrons to the source box, where gas molecules are ionised by collision. The positive ions thus produced, which are characteristic of the particular gas molecules, are then accelerated by an electric field between the source box and the slit plate and then travel along semi-circular paths under the influence of the magnetic field. Those of a particular charge-to-mass ratio will travel along an arc of a particular radius, usually about 10 mm, and pass through the collector slit to reach the collector plate. The positive-ion current to the collector is then electronically amplified and is a function of the number of ions and consequently the pressure.

The pressure recorded is in fact the partial pressure as it is the pressure exerted by the molecules of a particular gas. To obtain the pressure exerted by molecules of a different gas, the accelerating voltage must be altered to allow these positive ions to travel along an arc of the correct radius and thus pass through the collector slit. In practice, the accelerating voltages are calibrated in terms of the mass numbers of the elements and, besides manual setting, facilities are usually available for a continuous automatic sweep through the range of mass numbers, with the output from the collector and amplifier displayed on a chart recorder. The total pressure in the vacuum chamber is then the sum of the individual partial pressures. By turning to a light gas such as hydrogen or helium and using this gas as a search gas, a mass spectrometer becomes eminently suitable as a leak detector. The search gas may be applied to the region of a suspected leak by partially enclosing it in a plastic bag and filling the bag with the gas, or simply by using a small jet of the gas. With ultrahigh vacuum systems, however, what may be thought to be a leak may in fact be outgassing.

As with other gauges, the mass spectrometer does not have the same sensitivity to all gases and consequently the partial pressure of a particular gas must be calculated from equation 1.37 and *Table 1.4*. Commercially available instruments have a lower pressure limit of about 5×10^{-11} N m^{-2} (about 4×10^{-13} torr) and are free of X-ray limitations. The upper limit is set by physical deterioration of the filament and consequently the gauge should not be used above a pressure of about 10^{-2} N m^{-2} (about 10^{-4} torr). Owing to the high sensitivity of these instruments, contamination of the filament (especially by oil vapour) must be avoided otherwise the filament emission characteristics will be changed. For other than a 'clean' system, it is necessary to protect the mass spectrometer by a cold trap as, for example, was used with the McLeod gauge.

1.5 Pumping Systems

To pump down a chamber from atmospheric pressure to the required degree of vacuum generally requires different types of pumps to cover different pressure ranges. To obtain a high vacuum it is necessary to use first some type of roughing pump, generally a mechanical one, to bring the pressure down to about 1 N m^{-2} (approx. 10^{-2} torr), which brings the pressure into the range capable of being handled by the next stage pump, generally a diffusion pump. This type operates over the pressure range about 1 N m^{-2} (approx. 10^{-2} torr) to about 10^{-4} N m^{-2} (approx. 10^{-6} torr). When it comes into operation, the system acts in a sense as a two-stage one with the diffusion pump taking gas at low pressure from the chamber being evacuated and compressing it to a pressure that

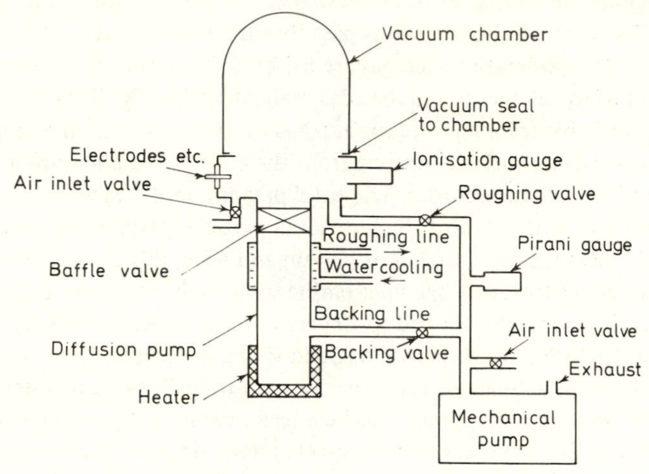

Figure 1.11. Arrangement of pumps and valves for a high-vacuum pumping system

can be pumped by the mechanical pump. The mechanical pump is then said to 'back' the diffusion pump.

Because a diffusion pump contains a hot working fluid, usually a silicone oil, it must not be exposed to air at atmospheric pressure. Thus a system of taps, or valves, is required as shown in *Figure 1.11.* Starting with all valves closed, the mechanical pump and Pirani gauge are switched on. The backing valve is then opened and the diffusion pump evacuated to a pressure of about 1 N m^{-2} (approx. 10^{-2} torr). The diffusion pump heater may then be switched on, as well as the water cooling required at the top end of the pump. Now the backing valve may be closed, the roughing valve opened, and the chamber evacuated to a pressure of about 1 N m^{-2} (approx. 10^{-2} torr).

Meanwhile the diffusion pump will have heated up and become operational so that the roughing valve may be closed, the backing valve opened again, the baffle valve opened, and the ionisation gauge switched on. Thus the diffusion pump is now pumping the chamber by moving gas from the chamber to the bottom of the diffusion pump, where it is at sufficient pressure to be removed by the mechanical backing pump. Provided it is remembered that the hot diffusion pump must not be exposed to the atmosphere, the operation of the valves follows in a logical manner.

To let air into the chamber again, it is necessary only to close the baffle valve, switch off the ionisation gauge, and admit air to the chamber. Pumping down again involves closing the air admittance valve

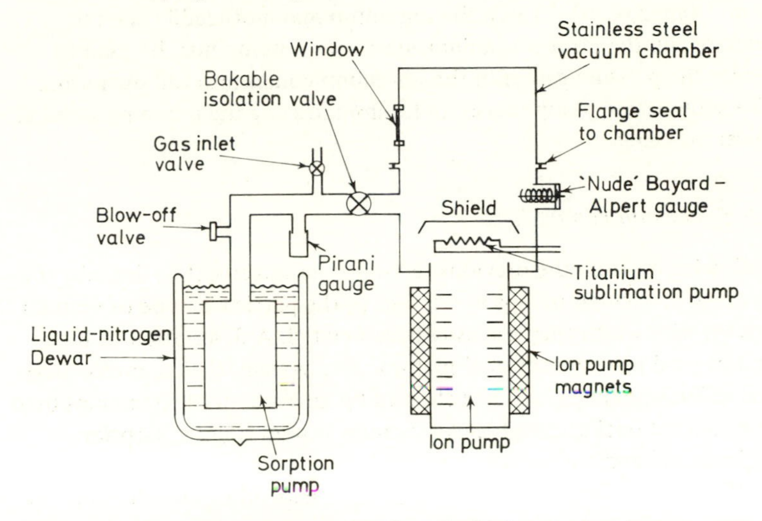

Figure 1.12. Arrangement of an ultrahigh vacuum system

and the backing valve, and roughing down again. To close down the system finally, the diffusion pump should be allowed to cool down whilst being backed by the mechanical pump. When cold it should be isolated by closing the backing valve and baffle valve, if not already closed. On turning off the mechanical pump, the air admittance valve should be opened immediately, otherwise oil may be sucked up from the pump and flood the system.

When an ultrahigh vacuum is required with pressures down to about 10^{-8} N m^{-2} (approx. 10^{-10} torr), then a different system of pumps is required. Although some improvement of the lower limit of a diffusion pump can be achieved with rather elaborate cold trapping techniques, it is still not very satisfactory. Since presumably an ultrahigh vacuum is required to avoid contamination of surfaces, it is inherently better to use an all-clean system where the pumps used do not require oil or mercury for their operation.

A typical arrangement for an ultrahigh vacuum pumping system is shown in *Figure 1.12*. A sorption pump is used to reduce the pressure of the complete system to about 10^{-1} N m^{-2} (approx. 10^{-3} torr) and the isolation valve is closed. The operating range of the ion pump is from about 10^{-1} N m^{-2} (about 10^{-3} torr) down to about 10^{-8} N m^{-2} (about 10^{-10} torr), although this can be improved upon with special techniques and system design. At the starting pressure of about 10^{-1} N m^{-2} (approx. 10^{-3} torr), the ion pump may not readily start to operate and therefore a titanium-sublimation pump may be used to bridge the pressure gap until the ion pump comes into full operation. The sublimation pump is also useful in increasing the pumping speed at lower pressures.

1.6 Vacuum Pumps

Following on from the previous section it is convenient to describe the working of vacuum pumps in relation to the particular type of vacuum system with which they are usually associated. A description of the pumps used in a conventional high-vacuum system, that is, mechanical and diffusion pumps, will be followed by an account of the pumps used in a conventional ultrahigh clean vacuum system, that is, sorption, sublimation, and ion pumps.

Mechanical pumps are of many designs, depending on different manufacturers and the speed required. A common type is the *vane type rotary pump*, shown diagrammatically in *Figure 1.13*. A rotor carrying a pair of spring-loaded sliding vanes is driven round inside a stator. Gas entering the pump will first occupy a volume marked A in the figure and will then be isolated by the vanes as the rotor rotates [*Figure 1.13 (b)*]. Further rotation compresses this gas into a smaller volume until its pressure is above atmospheric and it is forced out, escaping by lifting the flap valve. The pump is filled with oil to assist the sealing at the flap valve and at the sliding surfaces between the rotor, the vanes, and the stator. The lowest pressure attainable is limited by the back leakage of gas across the sliding oil seal between the rotor and stator, which limits the lowest pressure that can be reached to about 5×10^{-1} N m^{-2} (approx. 4×10^{-3} torr). A typical small pump for laboratory use may have an ultimate pressure of about 1 N m^{-2} (approx. 10^{-2} torr).

By coupling the outlet of one pump directly to the inlet of a second pump, a two-stage pump is formed. In this the pressure differential across the rotor–stator seal can be considerably reduced and, with it, back leakage of gas. This enables a lower ultimate pressure of about 10^{-2} N m^{-2} (approx. 10^{-4} torr) to be reached.

The pumping capacity of such a pump can be specified in two ways. In *Figure 1.13 (a)*, the rotating vane is about to isolate the gas in the

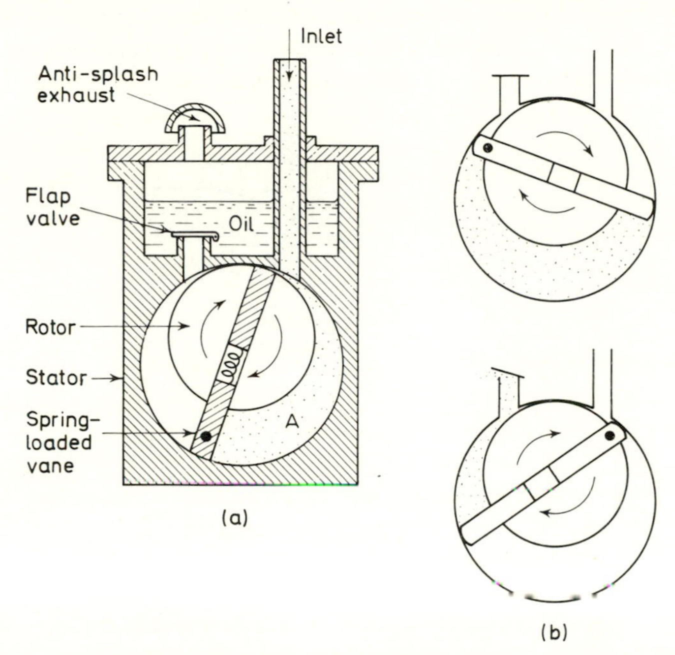

Figure 1.13. (a) Section through a vane type rotary pump, and (b) subsequent positions of the rotor showing the gas being isolated and compressed

volume marked A. This volume of gas will then be compressed, to be finally ejected. Since there are two vanes this cycle will occur twice for each revolution of the rotor. Thus, if V is the maximum volume of gas taken in, that is, the volume at A in *Figure 1.13 (a)*, the *displacement* of the pump is given by

$$\text{displacement} = 2nV \text{ m}^3 \text{ s}^{-1} \quad (1.38)$$

where n is the number of revolutions of the rotor per second, and the volume V is measured in cubic metres. These pumps are manufactured with displacements up to about 0·2 m^3 s^{-1}, with a displacement of perhaps 1×10^{-3} m^3 s^{-1} being a more common laboratory size. Obviously the displacement is equal to the speed of the pump at atmospheric pressure, since it is the volume of gas removed by the pump in unit time when the gas at the inlet is at atmospheric pressure.

The other way of specifying the capacity is to quote the actual pumping speed with pressure; for this type of pump the speed is not constant. As the pressure at the inlet is reduced by the operation of the pump, the speed decreases. This is due to back leakage, outgassing of the oil, and the fact that the volume swept out by the vanes is not decreased to zero – there is still a finite volume leading up to the

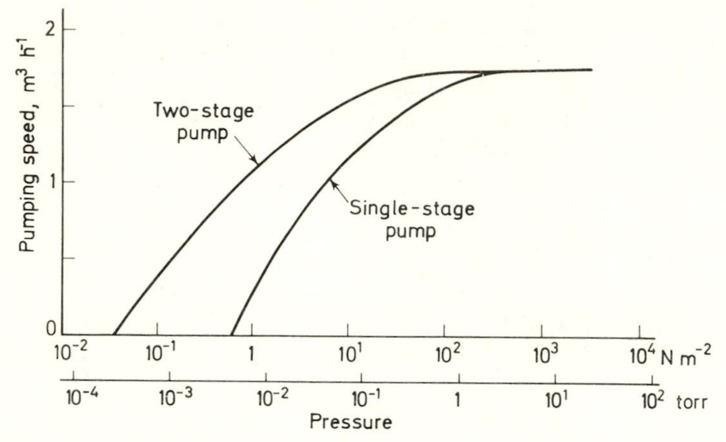

Figure 1.14. Variation of pumping speed with pressure for typical single- and two-stage vane type rotary vacuum pumps

exhaust flap valve. The variation of pumping speed with pressure is shown in *Figure 1.14* for typical vane type rotary pumps.

If the pump is used with an atmosphere which includes, for example, water vapour, then difficulties are encountered. If there is any free water in contact with the gas at the inlet, so that the gas is saturated, then the water vapour will exert a vapour pressure of about 2×10^3 N m^{-2} (15 torr) depending on the temperature of the gas at the inlet. In the pump, the gas, and with it the water vapour, will be compressed by an amount depending on the *compression ratio*. This is defined as the ratio of the maximum to minimum volumes swept out by the rotating vanes. Owing to this compression and to internal friction, there will be a rise in temperature and a consequent rise, perhaps tenfold, in the water vapour pressure within the pump. If now the compression ratio exceeds the ratio of the vapour pressures at the pump and inlet temperatures, that is, about 10 in this instance, then water will condense out in the exhaust cycle. For an inlet gas pressure of, say, 10^2 N m^{-2}, the compression ratio must exceed $1{\cdot}013 \times 10^5/10^2 \simeq 10^3$ for the pump to be able to compress the gas above atmospheric pressure to enable it to be discharged. Hence any efficient rotary vacuum pump will have a compression ratio that will cause condensation of water vapour. As free water it will be discharged to mix with the oil where,

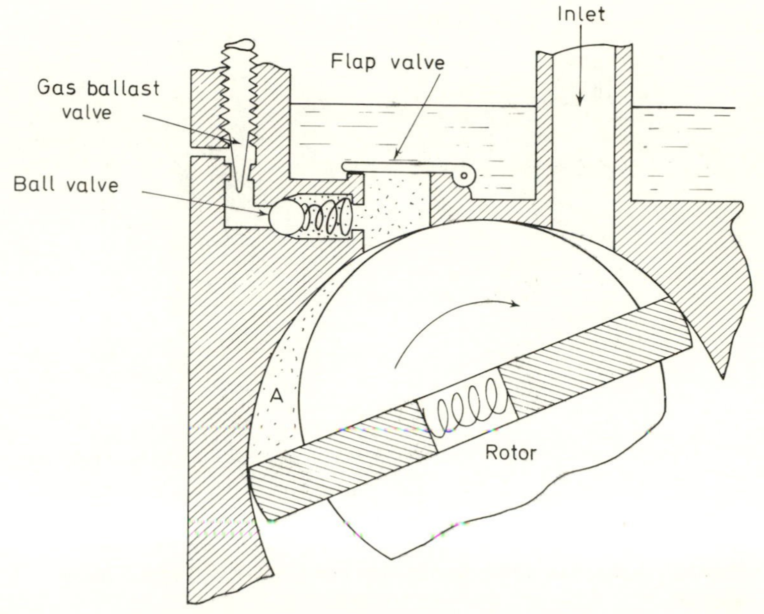

Figure 1.15. Part of section of vane type rotary pump showing the modification of the exhaust to provide gas ballasting

apart from causing serious deterioration of the oil, it will diffuse into the input side of the pump and return to vapour at the lower pressure. Thus it will continuously cycle between the exhaust and inlet sides of the pump and reduce the degree of vacuum obtainable, that is, the ultimate pressure will be increased.

To deal with this problem of condensable vapours, *gas ballasting* may be employed. This adversely affects the compression ratio but does allow the pumping of condensable vapours. The method is to allow dilution of the gas by admitting air through a non-return valve during the exhaust process. In *Figure 1.15* the gas and vapour in the volume marked A are at a pressure lower than atmospheric. Hence the ball valve will open and admit air at atmospheric pressure which dilutes the vapour–gas mixture. Further compression of this volume causes the ball valve to close and the exhaust flap valve to open. That is, the flap valve is opened and exhaust begins before the partial pressure of the water vapour reaches the saturation value. Since the pumping efficiency is impaired by the gas ballasting, the gas ballast flow is adjustable to a maximum of about 10% of the pump displacement. As the vapour is

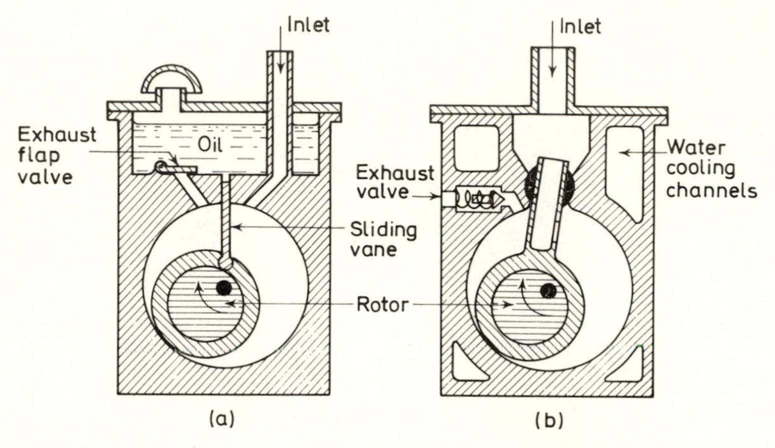

Figure 1.16. Sections through (a) a sliding vane type rotary pump, and (b) a Kinney rotating plunger pump

cleared from the system, the gas ballast valve can be progressively closed to enable the ultimate vacuum pressure of the pump to be reached.

Of the many designs of rotary vacuum pumps, the *sliding vane* type and the *rotating plunger*, or *Kinney pump*, shown diagrammatically in *Figure 1.16*, are in common use, with the rotating plunger type being used where large displacements are required. Gas ballasting facilities would normally be incorporated in these pumps, as well as two-stage versions.

Taking over from the lower limit of the rotary pumps, the diffusion pump can take the pressure from about 1 N m^{-2} (approx. 10^{-2} torr) down to about 10^{-4} N m^{-2} (approx. 10^{-6} torr), although by careful design and efficient cold trapping with liquid nitrogen several-orders lower pressure can be obtained. Generally, however, this is not warranted as ion pumps can be used with much greater efficiency at the lower pressures. An arrangement of a simple diffusion pump is shown diagrammatically in *Figure 1.17*. An electric heater causes the pump fluid, generally a silicone oil although mercury is still used for special applications, to boil and produce a copious supply of vapour. This vapour is ejected at high speeds from the jets. Molecules of gas are caught up in the vapour jet stream from the first and second jets shown in the figure and carried towards the bottom of the pump by

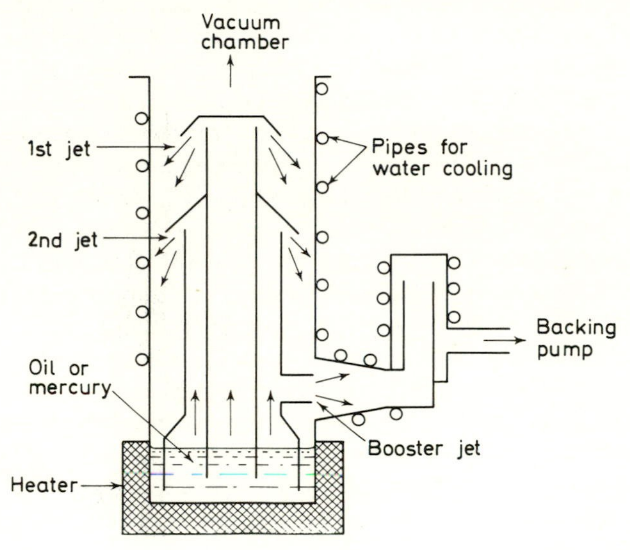

Figure 1.17. Operation of a diffusion pump shown diagrammatically

collisions with the fast moving heavy vapour molecules. Vapour contacting the cold walls is condensed and runs down to the boiler for recycling. The action of the jets causes gas molecules to accumulate at the bottom of the pump, with a consequent increase in pressure so that, for example, for an inlet pressure of 10^{-4} N m^{-2}, the pressure of the gas at the bottom of the diffusion pump may be raised to perhaps 1 N m^{-2} – a compression ratio of 10^4. At this higher pressure the gas can be effectively removed by a rotary pump.

The connection to this backing pump usually involves some form of cooled labyrinth to condense oil vapour from the diffusion pump and return it, otherwise it would be pumped away by the backing pump. Also commonly incorporated is a booster jet, as shown in the figure, to assist the backing pump and enable the diffusion pump to continue to work against high backing pressures. If, however, the backing pressure is allowed to rise above a certain value, the *critical backing pressure*, the vapour jet will be unable to carry gas molecules up the increased pressure gradient and the pump will cease to function. Hence the backing pump must be capable of removing gas from the base of the diffusion pump at a sufficient rate to maintain a pressure less than the critical value.

If some of the vapour molecules in the jet stream leave the jet with a velocity in a direction opposite to the main stream, then *back-streaming* occurs. The vapour molecules could also obtain this back velocity perhaps by collision with the walls of the pump, but either way there will be migration of some of the pump fluid molecules back into the chamber being evacuated. To prevent this a *baffle* is required to prevent an optical line-of-sight between the chamber and the diffusion pump. This may be a labyrinth path or more commonly is the flap valve that isolates the chamber from the pump.

Although silicone oil is the commonly used diffusion pump fluid, mercury still has many applications. A diffusion pump should not be exposed to the atmosphere while hot, otherwise the pump fluid will be damaged. This is especially so for oil, which will oxidise and decompose to form a black sludge of little use for pumping. Mercury too will oxidise if exposed while hot, but since it is constantly being redistilled it will recover its pumping efficiency provided it is not subjected to long and repeated exposure and, of course, it cannot decompose. The disadvantage of mercury is its high vapour pressure, so that, for pressures lower than about 10^{-1} N m^{-2} (approx. 10^{-3} torr), a liquid nitrogen cooled trap is required above the diffusion pump to condense the mercury vapour, whereas this is not necessary with silicone-oil filled diffusion pumps. Simple water cooling is then quite sufficient as silicone oils are now manufactured with vapour pressures to cover all requirements, with the choice only being limited by cost. Oils may also have a disadvantage when used with vacuum systems incorporating high voltages and electrical discharges. Under these conditions the oil is liable to be broken down into its constituent atoms. The older hydrocarbon vacuum oils formed carbon deposits, whereas the normally used silicone oils form insulating films of silicon or silicon oxide, which may be as troublesome as the conducting carbon deposits. Thus mercury has an advantage in these circumstances and similarly for use with, for example, gas analysis apparatus, since it does not decompose. Mercury in its turn has disadvantages when used with systems involving evaporated aluminium films as, unless elaborate cold trapping is used, the mercury vapour will destroy the newly formed aluminium film before it has had a chance to form a protective oxide layer.

As with rotary pumps, there are many variations in the design of diffusion pumps, which may have two, three, or four stages depending on the ultimate pressure and speed required and the backing pressure available. Jet designs may be modified to suit the different pump fluids and give greater pumping speeds over particular required pressure ranges.

The two types of pump so far described can in combination give high-vacuum conditions with pressures easily obtainable down to about 10^{-4} N m^{-2} (approx. 10^{-6} torr) and with difficulties down to a couple of orders lower. They do, however, require a pumping fluid, whether it be oil or mercury, with its consequent vapour pressure limitations and the risk of contamination of clean surfaces. Although there are mechanical pumps which are clean – for example the *turbo-molecular pump* as designed by Becker which works on the turbine principle and can attain pressures as low as 5×10^{-8} N m^{-2} (about 4×10^{-10} torr) – they generally have the disadvantages of being large and requiring very high rotational speeds with their inherent mechanical difficulties. Thus for preference a much simpler type of pump should be used when a clean system is required. The first in the series of commonly used clean pumps, which lead ultimately to ultrahigh vacuum conditions, is the *sorption pump*.

Sorption pumps can reduce the pressure in a system from atmospheric to about 10^{-1} N m^{-2} (approx. 10^{-3} torr), depending on the particular design and operating conditions. Apart from being clean, since they contain no oil, they have no moving parts and are silent. The pump is simply a cylindrical container, usually in stainless steel, containing what is known as a *molecular sieve* material. This is commonly calcium alumino-silicate in the form of pellets about 3 mm in diameter. During manufacture the molecular sieve material is heated to remove the water of hydration, leaving a very porous structure with uniform pores about 5×10^{-10} m in diameter. When cooled to liquid nitrogen temperature, gas molecules are adsorbed into the pellets and the pressure is rapidly reduced. The molecular sieve material is a poor thermal conductor and therefore the pump should be in the form of concentric cylinders to expose a large area to the liquid nitrogen, as in *Figure 1.18*. Liquid nitrogen is to be preferred to liquid air as the coolant, as with liquid air the nitrogen will boil-off first leaving a strong concentration of liquid oxygen with its inherent danger of fire or explosion. A quicker pump-down time is usually achieved using two sorption pumps – changing from the first to the second as the speed of the first falls away.

To reactivate a pump after use, it must be allowed to warm up. As this happens the gas molecules are released from the molecular sieve pellets and the pressure rises above atmospheric, to be released by a blow-off valve which can be simply a spring-loaded flap valve with a

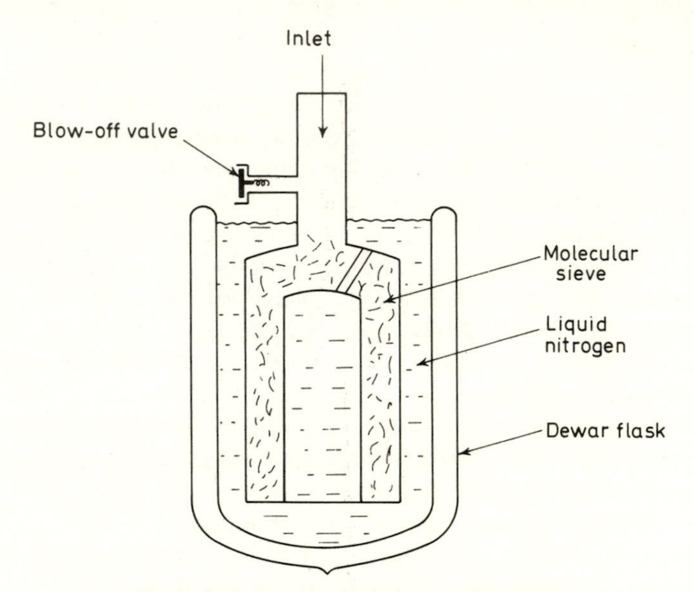

Figure 1.18. Section through a sorption pump

neoprene sealing ring. One disadvantage of this type of pump is that it can only adsorb a given quantity of gas during one cycle and is therefore unsuitable for pumping against a leak. A further disadvantage is that argon and hydrogen are pumped less efficiently than nitrogen and oxygen, leaving the vacuum chamber rich in argon and hydrogen. Argon is especially troublesome since this is pumped less efficiently by the ion pumps usually used in conjunction with sorption pumps, resulting in a vacuum chamber containing a relatively high proportion of argon as well as leading to argon instability with some types of ion pump. If, however, two sorption pumps are used, as has been suggested on the ground of speed of pumping, the air rushing into the first pre-cooled sorption pump tends to entrain the argon and hydrogen molecules so that the pressure, at least initially, is reduced non-selectively. If the change over to the second pump is made as soon as the main rush of air is finished, usually after several minutes, the argon and hydrogen have no chance to diffuse out again.

Sorption pumps also tend to sorb preferentially water vapour and to desorb it less readily at room temperature. Consequently it is occasionally necessary to reactivate completely the molecular sieve material by heating the pump at atmospheric pressure to about 300°C for several hours. Providing these pumps are treated reasonably, for example not contaminating the pellets with oil, and completely reactivated very occasionally as needs demand, then their life is virtually unlimited.

Having reduced the pressure in the vacuum chamber to about 10^{-1} N m^{-2} (about 10^{-3} torr), it is almost down to the firing pressure of the ion pump. To help bridge the small pressure gap and get the ion pump quickly to its more efficient operating pressures, a *titanium-sublimation pump* can be used. Before using this, however, it is necessary to close the valve isolating the sorption pumps from the vacuum chamber, otherwise the gas adsorbed by the molecular sieve material will be pumped back into the chamber.

The sublimation pump, or *getter pump* as it is sometimes called, works on the simple principle of evaporating titanium from a hot source and letting it condense onto a cold surface, for example, the walls of the chamber or a conveniently placed shield. The freshly deposited clean titanium atoms then pump by a gettering action, removing gas molecules from the free volume by chemically reacting with them. This gettering action will thus depend on the number and type of molecules and their sticking coefficients so that, for example, the pumping speed is fast for molecules such as CO, CO_2, H_2, etc., but virtually zero for the inert gases such as argon. Although designs of pumps are available for continuously feeding titanium wire onto a hot anvil, they tend on the whole to be troublesome to operate. A simpler system is to use a single tungsten wire, about 100 mm long, coated with titanium. By passing a heavy current through the wire the titanium is heated and caused to evaporate. Usually several of these titanium–tungsten wire elements are mounted side by side so that, when the titanium on one element is used up, the heating current can be switched to the next element.

Once the ion pump starts to operate, at a pressure perhaps a little below 10^{-1} N m^{-2} (approx. 10^{-3} torr), the sublimation pump is only required as an occasional boost to the ion pump, which will take the pressure down to about 10^{-8} N m^{-2} (approx. 10^{-10} torr), with a maximum pumping speed occurring over the range of about 10^{-3} to 10^{-6} N m^{-2} (approx. 10^{-5} to 10^{-8} torr). This lower limit of 10^{-8} N m^{-2} is mainly set by the cleanliness of the system and the slow

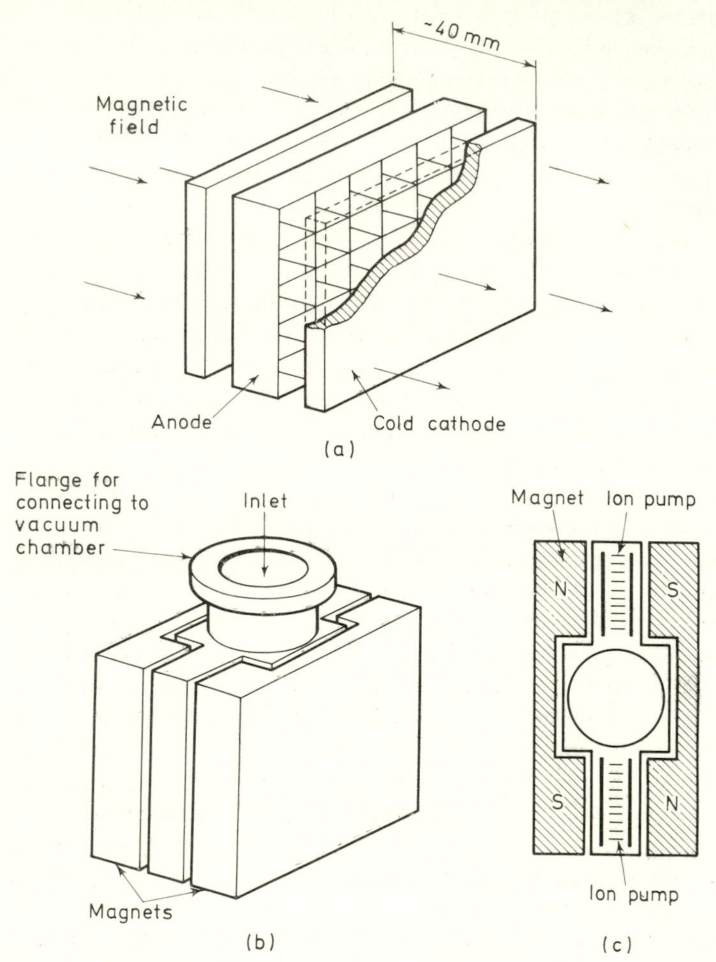

Figure 1.19. The arrangement of a multi-cell cold-cathode ion pump at (a) is shown in its practical arrangement at (b) which in turn is shown in section at (c)

outgassing of the surfaces. The principle of operation of the *ion pump* is the same as that of the cold-cathode ionisation gauge. Instead of the single cell anode between two plate cathodes of the ion gauge, many cells are arranged together in the pump as in *Figure 1.19*.

In the ion pump, or *Penning pump* as it is also called, the anode is usually of titanium but may be of some non-magnetic alloy, while the flat plate cathodes are of titanium, 1–3 mm thick. A voltage of up to about 10 kV is then applied between the anode and cathodes and a permanent magnet fixed to the outside of the pump casing to provide a magnetic flux density of about 0·2 T. This, as with the Penning gauge, is to provide a long spiral flight path for the ions and thus increase the chances of collisions with gas molecules to produce further ionisation. The pump casing is of stainless steel with the smaller pumps being sometimes of glass. Smaller pumps are usually mounted externally to the system and connected to it by a large bore tube, whereas the arrangement of *Figure 1.19 (b)* is compact and, more important, the conductance is not restricted by a tube.

The magnets, being mounted externally, are removable; this may be of assistance in installing the pump as the large ones can be quite heavy. Occasionally the pump may require baking to about 500°C for special cleaning purposes, when the magnets will have to be removed as they may otherwise be damaged. Care must be taken in replacing the magnets of the larger pumps as they will fly together and may damage themselves or the pump casing. Wooden wedges are therefore advisable to allow the magnets to be brought slowly into position. For routine use, however, with reasonably clean systems, baking to about 300°C is usually sufficient and it is not then necessary with most designs of pump to remove the magnets.

As with the Penning gauge, a discharge will be set up with the electrons moving towards the anode and the positive gas ions towards the cathodes. At the cathodes the arriving heavy positive ions will produce various effects. They will cause titanium atoms to be knocked off the cathode and ejected, a process termed *sputtering*; they will cause gas molecules adsorbed on the surface of the cathode to be ejected, that is, gas sputtering; and finally, they will cause electrons to be ejected from the cathode. These ejected electrons will then proceed on spiral paths towards the anode and will provide more ions by collision with gas molecules.

The main pumping action is due to the gettering action of the sputtered titanium which is deposited mainly on the large surface area of the anode. Here it pumps by forming a titanium compound with the gas molecules that collide with it. The gas molecules are thus removed from the active volume and consequently the pressure is reduced. It has been estimated that $\frac{1}{2}$–1 molecule of gas is removed from the active volume for each atom of titanium sputtered. The inert gases, which of course do not form compounds with titanium, are pumped by being adsorbed onto the electrode surfaces and buried by the arriving titanium atoms.

An ion pump has a number of advantages over, for example, a diffusion pump. Perhaps the most important of these is the cleanliness due to the absence of any pumping fluid, which also means that no baffles or cold traps are required. Since the principle of operation is virtually that of a Penning cold-cathode ionisation gauge, the discharge current is a function of the pressure. Thus with suitable calibration the pump can also be used as a pressure gauge although it may well record a slightly lower pressure than that existing in the chamber. To reach the ultimate pressure, it is necessary to drive off the adsorbed gas molecules from all the surfaces in the system, otherwise these will detach themselves over a long period and increase the pressure. To drive off these gas molecules, it is necessary to bake the complete vacuum system to about 300°C and hold at this temperature for some hours. By allowing the pump to cool quicker than the rest of the system, gas molecules will tend to be driven towards the pump and thence removed. Baking is usually carried out with an oven assembly that can be lowered over the complete vacuum system or with strip heaters fixed to the system at strategic places.

The lifetime of the pump is mainly dependent on the lifetime of the titanium cathode plates. With operation at a pressure of about 10^{-4} $N\,m^{-2}$ (approx. 10^{-6} torr) the lifetime may be perhaps 50000 h. At higher pressures the lifetime may be drastically reduced by the flaking-off of titanium from the anode, which may short out the electrodes, and by severe erosion of the cathodes. Consequently these pumps should not be used in systems where the pressure will normally be above about 10^{-3} $N\,m^{-2}$ (about 10^{-5} torr) or where the pressure is being repeatedly cycled. They should also not be used with, for example, a rotary backing pump without some form of cold trapping, otherwise the hydrocarbon vapour will poison the electrodes and seriously affect their sputtering properties.

A disadvantage with these pumps is the possibility of instability. This can be due to field-emission points forming on the electrodes followed by a surge of current in this localised region. The surge is sufficient to melt and destroy this point but another may form within seconds to give another current surge. The abruptness of the surges may cause difficulties with, say, the 10 kV d.c. power supply and may lead to failure of components within the power supply. Growing field-emission points can be removed by judicious banging of the pump casing to shake them loose or by temporarily using a power supply giving about 50% over the operating pump voltage to burn-off growing points.

The other source of instability which is not easily curable is *argon instability*. As has already been said, the inert gases of which argon is the most common will be pumped mainly by adsorption on the electrode surfaces. Argon adsorbed onto the cathode surface, however,

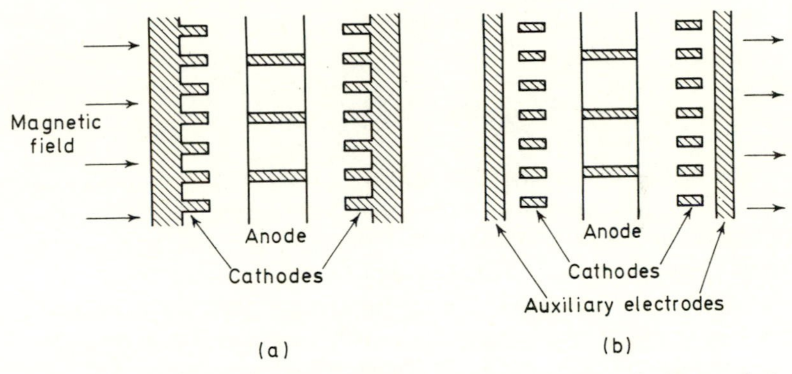

Figure 1.20. Prevention of argon instability (a) by using slotted cathodes in a diode ion pump or (b) by using a triode pump

will be released again by gas sputtering processes owing to arriving positive ions of other gases. The argon atoms tend to be released together to form sudden pulses of pressure, up to perhaps 10^{-2} N m^{-2} (or 10^{-4} torr), occurring at regular intervals which may be of the order of minutes or even hours. The instability is especially noticeable if the pump has been working against a leak for some hours. To reduce argon instability, the cathode plates may be slotted as in *Figure 1.20 (a)*. With this design the incident positive gas ions tend to strike the sides of the slots owing to the action of the magnetic field, so that the sputtering of titanium is also mainly from the sides. The inert gases, however, will tend to be adsorbed on the bases of the slots and will thus not be sputtered by the incident gas molecules. They will also become buried by sputtered titanium.

An alternative, but more costly arrangement mainly owing to the power supply, is to use a triode pump rather than the diode one so far described. This is shown in section in *Figure 1.20 (b)* where the cathodes are now perforated and auxiliary electrodes are used, also of titanium. The auxiliary electrodes are maintained at a positive potential which is less than the anode potential. Sputtering of titanium will thus mainly take place from the perforated cathode because, owing to the action of the magnetic field, the sputtering ions strike at large angles of incidence, whereas the ions striking the auxiliary electrodes will do so at small angles of incidence and with little energy due to the electrode potential. However, inert gas adsorption will be mainly on the comparatively large area auxiliary electrodes, with consequently less chance of being freed by the arriving gas ions. They will also be buried by sputtered titanium from the cathode.

1.7 Vacuum Components

The constructional material of the vacuum system is determined mainly by the ultimate vacuum pressure requirements. All-glass systems are used for specialised processes and for example even all-glass diffusion pumps may be employed, but generally a metal system is used, with the possibility of a glass bell-jar for the vacuum chamber. The most suitable metal for piping in a high-vacuum system is copper. Copper piping is seamless, available in a range of diameters, and is easily bent and soldered. Brass is often used for the base plate of the vacuum chamber but owing to the zinc in brass it is not suitable for use in systems where elevated temperatures may be encountered. Having been heated and lost zinc, the brass may well be porous and for this reason it is often nickel plated before use. For ultrahigh vacuum systems which require baking, possibly occasionally to 500°C, the constructional material must retain its mechanical strength while evacuated and, more important, must have a low vapour pressure. It must also be impervious to the diffusion of gases through it. For these reasons, stainless steel has become the most popular material even though it is mechanically a relatively difficult material to work and must be argon arc welded to obtain good vacuum tight joints.

Synthetic rubber, usually neoprene or Viton, can be used for seals between demountable components and in valves not subject to elevated temperatures and where its vapour pressure is acceptable as in high-vacuum systems. For ultrahigh vacuum systems the sealing material must be able to withstand baking and must have a sufficiently low vapour pressure. This limits the sealing material to a metal, with copper or gold being the usual materials. The use of these materials for sealing is illustrated in *Figure 1.21*, which shows one type of O-ring seal for joining two pipes for high-vacuum use and a knife edge and copper seal

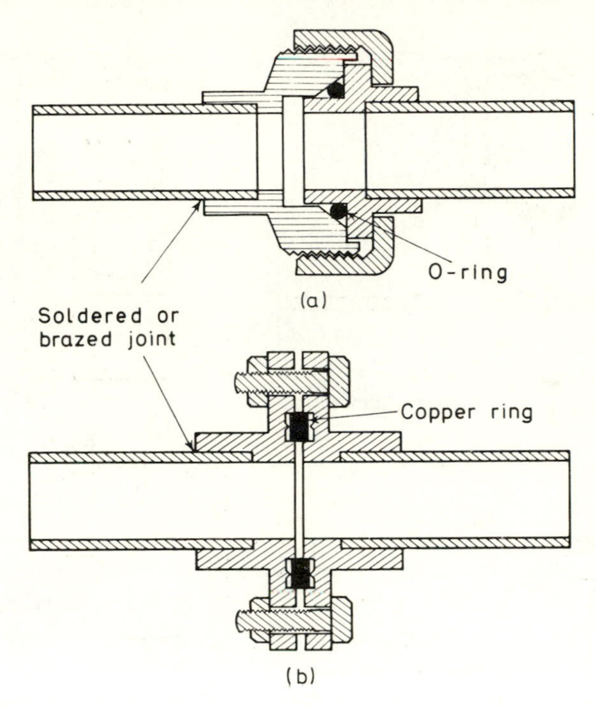

Figure 1.21. Section through (a) a demountable O-ring tube coupling suitable for high-vacuum use, and (b) a knife-edge copper ring coupling for ultrahigh vacuum use

for ultrahigh vacuum use, and in *Figure 1.22*, which shows diagrammatically sections through typical designs of valves, again using similar sealing techniques.

In addition an L-section gasket of synthetic rubber is used for sealing a bell-jar chamber to a base plate in a high-vacuum system and usually a gold wire seal for sealing the chamber cover in an ultrahigh vacuum system. A gold wire seal is simply a ring of annealed gold wire squashed between two flanges as the flanges are bolted together. When using O-ring seals it is important that the correct dimensioned groove is used to retain the ring. Sealing is here due to squashing the ring to about 80% of its original diameter and allowing the correct space in the groove for the displaced volume. A smear of silicone grease around the O-ring is usually used to assist the seal.

A common need is to provide a rotary motion into a vacuum chamber. This can be done using magnetic coupling or, if something

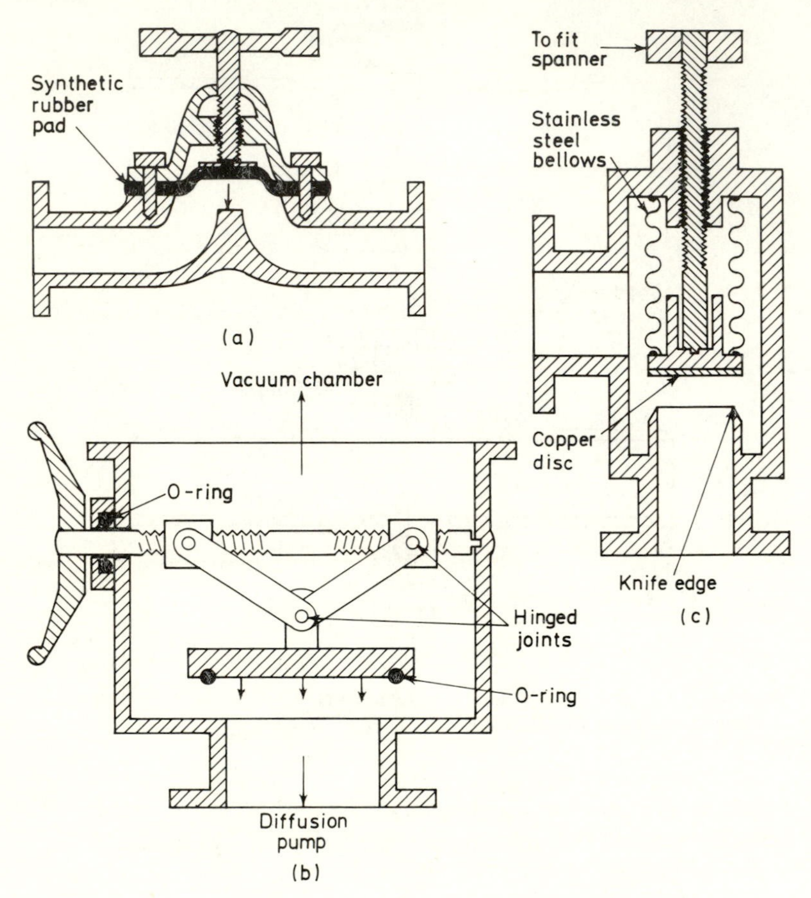

Figure 1.22. Sections through a Saunder's valve (a) and a baffle valve (b) for use in high-vacuum systems, and a bakable valve (c) for ultrahigh vacuum use

more positive is required, by using a rotary seal as in *Figure 1.23*. The Wilson seal shown diagrammatically in *Figure 1.23 (a)* is suitable for high-vacuum use and has the added advantage that it can give an in-and-out motion, provided the shaft is well polished and free from scratches. The seal also depends on a generous coating of silicone grease for

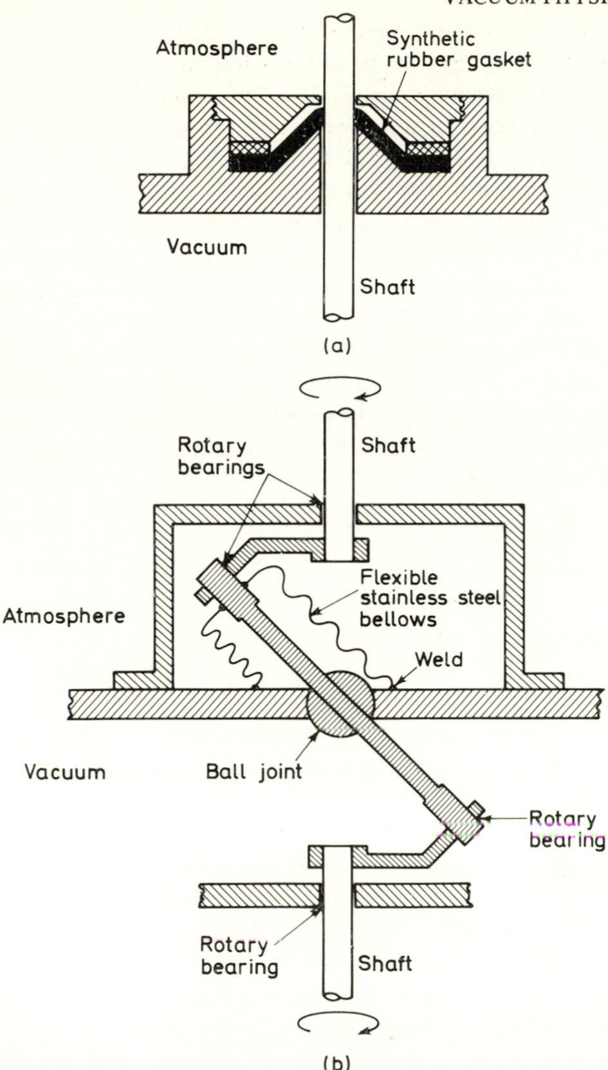

Figure 1.23. Rotary motion seals for (a) high vacuum, and (b) ultrahigh vacuum use

efficient operation and it is not suitable for rapid motions. In *Figure 1.23 (b)* is shown, again diagrammatically, a 'wobble' drive suitable for giving rotary motion into an ultrahigh vacuum system.

Little has been said concerning cleanliness in a vacuum system but obviously anything that has a higher vapour pressure than the ultimate pressure required in the system must be avoided. Thus, for example, greasy finger-prints must not be left in a system in which it is hoped to attain ultrahigh vacuum pressures. Likewise, roughly machined surfaces and blind screw holes must be avoided otherwise gas molecules swaged into the surface or adsorbed will steadily leak into the active volume to raise the pressure, as will the gas molecules trapped under the screws. For the very lowest pressures, surfaces must be electrolytically cleaned and the use of any hydrocarbon solvents avoided. For high-vacuum conditions such stringent care is not required, nevertheless care must be taken if a sufficiently low pressure is to be achieved. One method of cleaning in a high vacuum system is to employ a gas discharge produced by a potential of several kilovolts between electrodes placed in the vacuum chamber. By holding the pressure at a suitable value a visible glow discharge can be maintained for 10 min or so, which provides an effective final cleaning.

APPLICATIONS

1.8 Vacuum Evaporation and Coating

As was mentioned in the introduction to this chapter, if a metal or for that matter a range of other substances is heated in a vacuum, it can be evaporated and will coat everything 'seen' by the hot source. One of the easiest materials to evaporate is aluminium, which is used for both decorative and optical purposes. Any high-vacuum system is quite satisfactory for this operation providing an oil and not a mercury diffusion pump is used.

Using tungsten wire as the hot source, a helix is made by winding the wire several times around a pencil, for example, and then mounting between the low-tension electrodes in the vacuum chamber. Only one of the electrodes is normally insulated, the other being connected to the earthed base plate. To clean the tungsten wire prior to the evaporation process, it is necessary to heat it momentarily to bright red heat whilst under vacuum by passing an electric current. After returning the system to atmospheric pressure and removing the bell-jar, pieces of aluminium wire are hung on the loops of the helix, as in *Figure 1.24 (a)*, and pinched firmly in place, otherwise they may be shaken off by the vibration of the rotary pump. If a point evaporation source is required, then the tungsten wire should be bent into a 'hair-pin', as in *Figure 1.24 (b)*. Both tungsten and high-purity aluminium wires are readily available commercially. Tungsten wire of 0·5 mm diameter is generally

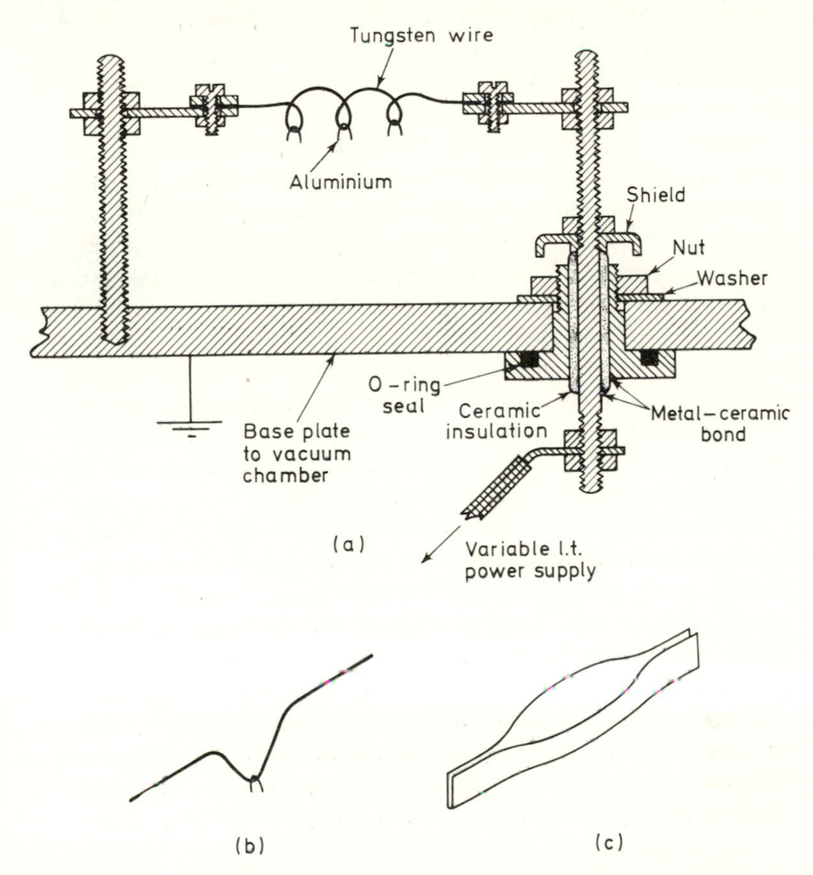

Figure 1.24. Arrangement of heating source for the vacuum evaporation of aluminium (a) *and alternative 'hair-pin'* (b) *and 'boat'-* (c) *shaped sources*

satisfactory, although thicker or multiple stranded wire is more suitable for industrial applications, while the use of multistrand wire also reduces any tendency towards 'spitting' by the molten aluminium.

Now the system should be pumped down again to a pressure of about 10^{-2} N m^{-2} (approx. 10^{-4} torr) or lower, after which the temperature of the tungsten wire may be raised slowly, when the aluminium will be seen to melt into globules and 'wet' the tungsten wire. On heating the wire to a bright red, above about 1100°C, the aluminium will freely evaporate and coat everything within its optical line-of-sight. Any dirt or contamination on the surface of the aluminium will tend to be evaporated first. Therefore a pre-evaporation run should be made without the object to be coated being in the chamber or preferably, since it avoids another evacuation cycle, a shutter should be used to shield the object from the first few seconds of the evaporation. This shielding from the initial evaporation is especially important for mirrors which will be used by viewing from the glass side. For front surface mirrors, this precaution and the pre-cleaning by heating the tungsten may not be necessary, although now it is advisable that the evaporation should be stopped before all the aluminium has been evaporated from the tungsten source.

Although more aluminium can be added to the tungsten wire element for further evaporations, in practice the life of the tungsten element is quite short. A chemical reaction takes place between the aluminium and tungsten when hot, causing the tungsten to become extremely brittle and fragile. Alternatively, aluminium wire may be fed continuously onto a heated tungsten anvil.

As has already been said, all objects in general may be coated but it should be remembered that the evaporated atoms travel along rectilinear paths from the source and therefore it may be necessary to rotate the object in a suitable manner to ensure even coverage. Conversely, by taking into account the inverse square law effect, uneven coverage may be utilised. Thus, by suitably positioning the source relative to a concave mirror and rotating the mirror during the evaporation cycle, a concave spherical mirror may be parabolised. Masks may also be used to delineate specific areas to be coated, but evaporation through very narrow slits or pin-holes is not possible owing to outgassing effects from the edges of the masks.

To obtain a strongly adherent and stable coating, it is most important, however, that the surface to be coated should be clean and especially free from grease. For example, glass should be cleaned first with soap and water or, better still, mechanically cleaned by wiping over lightly with a water paste of fine polishing rouge, for example cerium oxide. Care is essential in this to see that no extraneous grit is allowed to scratch the surface. The glass should then be washed clean in distilled water, and dried by sluicing liberally with alcohol and allowing to drain while standing on edge. If the aluminium coating is unsatisfactory for any reason, it is easily removed with a dilute solution of sodium hydroxide. Incidentally, since aluminium is so easily removable, it is useful to perform an aluminium evaporation to coat the inside of the vacuum chamber before an evaporation involving a more chemically inert material such as gold to give ease of cleaning.

On removal from the vacuum chamber, the freshly deposited aluminium coating should be allowed to stand protected from dust for several hours to harden. A freshly deposited aluminium film feels 'sticky' if lightly rubbed and requires time to consolidate and form a thick oxide coating. The coating may be further hardened by burnishing with a soft cloth or tissue, although it is extremely difficult to avoid covering the surface with fine scratches. A better protection, however, is to evaporate a layer of quartz over the aluminium.

The above method of evaporation for aluminium is generally applicable to other metals providing they 'wet' tungsten. For metals and other materials that do not wet tungsten and are therefore liable to drop off when molten, other refractory metals such as molybdenum, tantulum, niobium, or even chromel or nickel may be tried in wire form or, alternatively, by folding a thin strip, may be formed into a boat-shaped heater. Some molten metals, such as iron, tend to be very chemically reactive with the refractory metals but may, however, be evaporated by being allowed to act as their own heating source. Thus, by passing a current through a piece of iron wire to heat it to just below its melting point, the iron will be caused to evaporate. This method requires very careful control of the current otherwise the wire will melt. A more satisfactory method, giving a faster evaporation rate, is to melt the metal in a crucible which in turn can be heated by a tungsten wire element. The crucible can be of silica wound round with the tungsten wire and then coated with an alumina cement. The cement is necessary to achieve good thermal contact with the crucible, otherwise the heating element will burn out. Alternatively, the tungsten heater wire may be wound into a conical coil to form a basket and then coated with alumina cement, which itself becomes the crucible. It is essential that the alumina cement be allowed to dry out slowly and completely when it may then be very gently heated to red heat to make it set. Nevertheless, it will still be fragile and requires careful handling.

Where it is important that there should be no possibility of contamination of the molten metal by the crucible or heating element, recourse may be made to an electron gun method of melting. Electron guns may be obtained commercially and comprise a hot wire filament to give a copious supply of electrons. These electrons are focused by a suitable configuration of electrode plates to a spot about 20 mm from the end of the gun. Owing to the very local intense heat spot, the material to be melted can act as its own crucible with solid material surrounding the molten area. This method is, however, expensive mainly due to the required kilovolt power supply and is only effective with electrically conducting materials, otherwise the surface charge built up deflects and defocuses the electron beam. Care must also be taken with the crucible supports, as their thermal expansion can move the crucible out of focus. Metals which sublime, such as chromium or titanium, are most easily evaporated by electroplating them onto tungsten wire which acts as the heater element.

The evaporation of non-metals generally requires a boat-shaped heating source. Some materials tend to splash and spit when molten or, like magnesium fluoride, which does not melt but sublimes and is used in the 'blooming' of lenses, become violently agitated on heating. These materials are best retained by covering the evaporating boat with a lid containing a hole. Experiment best decides the size of the hole, which should be large enough to allow a copious evaporation but small enough to retain most of the supply of material being heated.

One further method of evaporation deserves mention and this is the method of flash evaporation. In this the material to be evaporated should be in the form of a fine powder. It is allowed to trickle slowly onto a very hot anvil where it is instantly melted and evaporated. This method has the advantages that contact with the heating element is at a minimum, and with it the likelihood of chemical contamination, and also by using a mixture of two powders an alloy can be evaporated. The evaporation of an alloy from a boat or crucible usually presents difficulties as the different melting temperatures result in preferential evaporation of the lower melting point material.

Little yet has been said of the uses of vacuum coating but nevertheless they are manifoldly diverse. Mention has already been made of the optical usage for the production of mirrors and the blooming of lenses. In addition, semi-reflecting coatings are made by depositing aluminium films sufficiently thin to be transparent. Similarly, by depositing alternate dielectric and metallic films, multilayer interference filters may be made. Apart from the various research uses, in studies requiring clean surfaces for example, mention may also be made of the uses in the electronics industry, such as in the production of microelectronic devices where thin film resistors are deposited and alternate layers of aluminium and dielectric can be evaporated to form thin film capacitors. Also, by depositing a metal layer onto a semiconductor and heat treating to give a controlled diffusion of the metal into the semiconductor, transistor devices are now manufactured.

1.9 Sputtering

A different method of producing coatings is by sputtering, which has the advantage that, for example, the structure of the deposited film may be modified due to the increased number of deposition parameters. Thus a transparent metal oxide film may be deposited with a particularly fine structure which is of use in optical beam splitters since the light absorption will be reduced. Mention has already been made of the effect of positive gas ion bombardment of a cathode in the operation of an ion pump; it is this same process that is utilised for sputter deposition.

The metal to be sputtered forms the cathode, which can be a wire or in sheet form although a sheet generally proves more satisfactory in practice, while the anode, normally at earth potential, can be the object to be coated or as is more usual a metal plate parallel to the cathode. A power supply is required capable of maintaining a potential of several kilovolts between the electrodes, with a current up to perhaps half an ampere. The pressure must be adjusted to produce a glow discharge in which positive gas ions will bombard the cathode causing atoms of the cathode material to be ejected. These sputtered atoms will then condense on the surrounding surfaces, including the objects to be coated, and a film will be slowly built up. To prevent wasteful sputtering of atoms from the far side of the cathode, it may be supported closer to any neighbouring earthed metal surfaces than the length of the cathode dark space, hence suppressing the glow discharge on this side.

The rate of sputtering from the cathode depends on the voltage and current density, which in turn are dependent upon the gas density. Since the cathode, and with it the residual gas, becomes hot during the sputtering process, it is necessary to adjust the pressure continually to maintain the optimum gas density and sputtering conditions. The cathode is sometimes water cooled to assist in attaining stable conditions. To maintain the glow discharge necessary for a high sputtering rate, the pressure is usually in the region of 1 to 10 $N\,m^{-2}$ (approx.

10^{-2} to 10^{-1} torr) and, since the sputtering rate is found to increase with the atomic weight of the gas in the system, the atmosphere is usually one of argon. This also has the advantage of preventing undesirable chemical reactions with the newly deposited film. The vacuum system should therefore be evacuated to as low a vacuum as is practicable to outgas the chamber and objects to be coated, and then argon leaked into the system. Leaving the diffusion pump in operation, the leak rate of argon can be adjusted to give a suitable pressure for sputtering. By using an atmosphere of oxygen and argon, 'reactive' sputtering may be performed in which a pure metal cathode may be sputtered to produce a metal-oxide film.

A modification to this method of sputtering allows a non-electrically conducting material to be sputtered. With the arrangement already described, the material to be sputtered forms the cathode; with a dielectric material, however, this is impossible. The method therefore is to mount the dielectric in front of a metal backing plate. To this backing plate is applied a radio frequency (r.f.) voltage, which induces an r.f. voltage on the front surface of the dielectric target. This enables electrons and positive ions in turn to be attracted towards the dielectric target but, since electrons are more mobile than positive ions, more electrons reach the target during the positive half cycle than positive ions during the negative half cycle. Thus the dielectric target becomes negatively biased, remaining negative for a longer time than it is positive during a cycle. Positive ions are therefore attracted during the negatively charged period to produce sputtering and electrons attracted during the remaining positive period of the cycle to neutralise the accumulated positive ions, hence preventing a permanent surface charge being built up. To allow sufficient negative bias to build up and to give adequate time for electrons to arrive at the target and neutralise the accumulated positive ions, a frequency of several megahertz is generally most satisfactory. An r.f. voltage of several kilovolts is required to give a satisfactory rate of sputtering, coupled with an argon pressure of about 10^{-1} to 3 N m^{-2} (approx. 10^{-3} to 2×10^{-2} torr). Under these conditions a plasma is formed, and may be sustained by the r.f. voltage applied between the target electrode and the surrounding earthed chamber and work table or by the use of a hot filament cathode and a side anode, termed a triode arrangement. Distances from the target to the object to be coated are typically 25–100 mm. It is also usual to apply an axial magnetic field of up to about 10^{-2} T.

Uses of r.f. sputtering have increased rapidly since about 1965 and the method is now used for metals and semiconductors as well as insulators.

1.10 Other Applications

Apart from individual specialist uses of vacuum systems and techniques for research purposes, there are numerous industrial applications. Some of these have already been mentioned but, as well as the obvious uses such as in the production of television tubes and valves, the following may serve as additional examples.

Increasing use of vacuum techniques is now being made in the food industry, for example, the freeze drying of foods is now an important method of preservation. Here the material is first quick frozen to retain its structure and then dried under vacuum with ice passing direct to vapour.

In the chemical industry, use is made of the fact that liquids under vacuum boil at reduced temperatures, enabling substances of high molecular weight that would decompose or oxidise at higher temperatures to be distilled. In metallurgy, molten metals may be degassed under vacuum so that on cooling a homogeneous solid without occlusions at the grain boundaries is formed.

Finally vacuum impregnation, whether it be impregnating a wooden post with creosote or transformer windings with epoxy resin, may be mentioned as another industrial use. In this the article to be impregnated is put into a vacuum chamber which is then evacuated. Whilst still under vacuum, the object is dropped into the impregnating liquid. On returning to atmospheric pressure, the impregnating liquid is forced into all the pores.

Two

PHOTOGRAPHY AND OPTICAL INSTRUMENTS

PHOTOGRAPHY

2.1 Cameras

A camera consists basically of a lens that focuses an image of an object onto a photographic plate or film, and a shutter that controls the time for which the image is allowed to fall onto the film. Providing that the object is stationary and a long exposure is practicable, a perfectly good photograph may be obtained using only a simple lens system and a small entrance pupil. However, such a camera is not very versatile so most modern cameras have lenses corrected to give a sufficiently good image with a large entrance pupil, thus allowing enough light to enter in a short exposure time. With a sufficiently short exposure time, the photography of rapidly moving objects is no problem.

To obtain short exposure times it is necessary that the camera lens should have a large aperture, which considerably increases the problem of designing a lens that is sufficiently free from aberrations. With the general use of colour photography, the lens must obviously be substantially free from *chromatic aberration*, that is, light of different wavelengths must be brought to a common focus. By combining positive and negative lenses made from different dispersion glasses, it is possible to obtain perfect chromatic correction for two wavelengths, which for photographic purposes are usually chosen to be blue and green. The lens should also be corrected for *spherical aberration*, that is, parallel rays entering through different radial zones of the lens should come to a common focus and, since the image is to be recorded on a flat film, *curvature of field* must also be reduced to a minimum. Associated with the curvature of field problem is that of *astigmatism*. This is due to rays from an off-axis point object taking skew paths through the lens system as in *Figure 2.1* in which various rays from an off-axis point A are shown refracted. Rays in the tangential plane, defined by RSA, are brought to a focus at T, whilst rays through the sagittal plane, defined by PQA, are brought to a focus at S. If a screen is passed through the focusing region, the images seen are successively an ellipse, a line, an ellipse, a circle, an ellipse, a line, and an ellipse; a point image is never

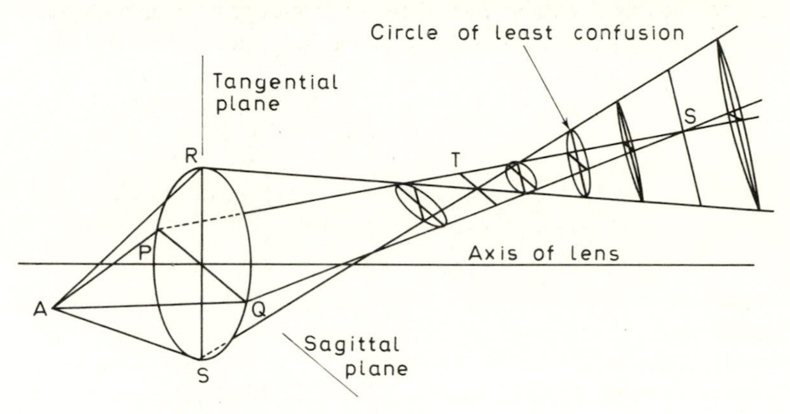

Figure 2.1. Ray paths through an astigmatic lens

formed. The best representation of the object that can be achieved is a circular disc, termed the circle of least confusion.

If the positions of the images T and S are determined for a large number of off-axis points such as A, that is, a large plane object normal to the axis, it is found that the loci of T and S are two ellipsoidal surfaces, termed the tangential and sagittal image surfaces respectively. The amount of astigmatism, the astigmatic difference, is the distance between these two surfaces as measured along a chief ray – that is, the ray drawn through the object point in question and the centre of the entrance pupil, which in a simple case would be the centre of the lens. More correctly, the entrance pupil may be defined as the image formed by all the lenses preceding it of the aperture stop that limits the extreme marginal rays through an optical system, whereas the exit pupil is the image of this aperture formed by all the lenses following it, that is, the exit pupil is the image of the entrance pupil.

It should also be noted that the astigmatism referred to is not the same as the defect of the eye referred to as astigmatism. Astigmatism of the eye is due to the eye lens having what is effectively a cylindrical component. This results in different magnifications in different radial directions so that a point object will be imaged as a line, even when it is on the axis of the eye lens.

To flatten the image field so that it is uniformly in focus over the whole film area, a stop is used to restrict the ray paths through the outer zones of the lens. With suitable positioning, the stop can cause the tangential and sagittal surfaces to be curved in opposite directions and, by the use of suitable glasses, the surfaces may be made to cross over making the astigmatic difference zero for points near to the edge

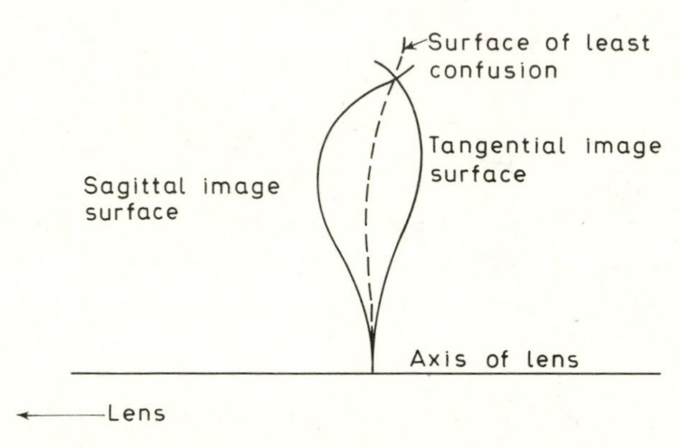

Figure 2.2. Reduction of astigmatic difference by crossed astigmatic image surfaces

of the field, as in *Figure 2.2*. Thus a lens may be made, termed an anastigmat, in which an approximately flat field is obtained with a minimum of astigmatism.

A further aberration of especial importance is *distortion*, which results if the lateral magnification is not constant over the image plane. If the magnification increases with distance from the axis, the distortion is said to be positive and a square object will be distorted into a 'pin-cushion'-shaped image. Conversely, with a decreasing magnification, negative distortion and a 'barrel'-shape results. Distortion is a particularly noticeable aberration in photography and its complete elimination is therefore desirable. Since it is more objectionable than a slightly unsharp image, the tolerances on spherical aberration and astigmatism are usually slightly reduced in order to give a greater latitude in design and allow better distortion correction to be obtained.

The remaining aberration, *coma*, is also of importance. This arises again when rays from an off-axis object pass through the lens and is due to the magnification being different for rays through different radial zones. However, it usually follows in practice that, if the spherical aberration has been reduced to within a satisfactory limit, then so also has the coma. It must also be remembered that it is theoretically impossible to reduce all aberrations simultaneously to zero and a compromise must always result. It is found that a satisfactory compromise may be obtained with a near-symmetrical arrangement, as in *Figure 2.3*, and the greater the number of components the closer to the

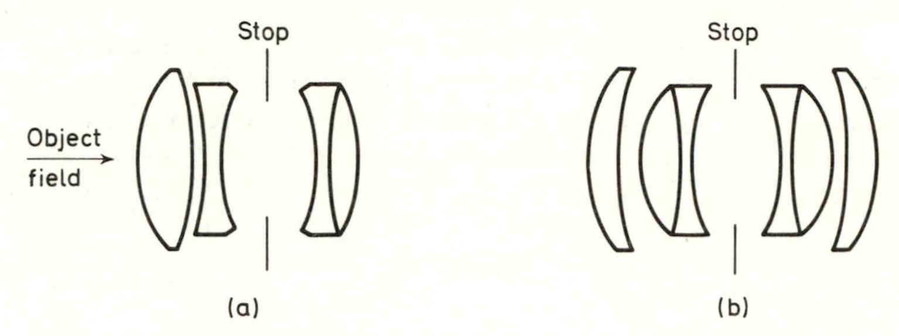

Figure 2.3. Typical four-element triplet (a) and six-element four-component (b) photographic lens systems

ideal will be that compromise. Wherever there is any intention of enlarging from the negative, a three-element anastigmat is generally a minimum requirement.

The stop already referred to is in fact a variable aperture diaphragm and is also used to control the amount of light entering through the lens. It is termed the *aperture stop*. The time taken for the photographic image to be formed is a function of the amount of light falling on the photographic film; the ability of the lens to collect light is its *speed*. The amount of light entering the lens is proportional to the luminance of the object being photographed and to the solid angle subtended by the aperture stop at the source. *Luminance* is the number of lumens emitted per steradian (unit solid angle) by unit area of the source or object. The solid angle is the angle subtended by any area A on the surface of a sphere of radius r and is by definition equal to A/r^2 steradians. Thus the solid angle subtended at the source or object by the aperture stop is $\frac{1}{4}\pi a^2/x^2$, where a is the diameter of the aperture stop and x is the distance to the object. Therefore the amount of light entering the lens from the object is proportional to a^2/x^2. How much of this light falls on unit area of the actual image formed at the film varies inversely as its area, which varies inversely as the square of the lateral magnification – that is, as $(x/x')^2$, where x' is the image distance. So the illumination of the image, that is, the number of lumens falling on unit area of the photographic film is proportional to

$$\left(\frac{a}{x}\right)^2 \left(\frac{x}{x'}\right)^2 = \left(\frac{a}{x'}\right)^2 \tag{2.1}$$

For a distant object, the image distance becomes approximately equal to the focal length, f, of the lens and the illumination then becomes proportional to $(a/f)^2$.

The number a/f is therefore useful in determining the necessary exposure time to build up the correct density photographic image. However, since a/f is usually a fractional number, the reciprocal is normally quoted; this is called the *focal ratio* or *f-number* and is defined as

$$\text{f-number} = \frac{f}{a} \tag{2.2}$$

where f is the focal length of the lens and a is the diameter of the entrance pupil. The illumination of the image is therefore proportional to $1/(\text{f-number})^2$. For example, a lens of focal length 45 mm and linear aperture 15 mm is an f/3 lens. This means that an f/3 lens is 'faster' than an f/3 lens stopped down to f/4 in the ratio $4^2 : 3^2$. The f/3 lens is thus $16/9 = 1{\cdot}78$ times faster than the f/4 lens; putting it another way, the exposure time required to photograph an object would have to be increased 1·78 times if an f/4 lens were used instead of the f/3. The adjustable aperture stop is normally calibrated to give known f-values. For example, a typical series of stop numbers would be

f/ 22 16 11 8 5·6 4 2·8

That is, the corresponding illuminations of the image would be as

$$\frac{1}{22^2} : \frac{1}{16^2} : \frac{1}{11^2} : \frac{1}{8^2} : \frac{1}{5{\cdot}6^2} : \frac{1}{4^2} : \frac{1}{2{\cdot}8^2}$$

i.e.

$$\frac{1}{484} : \frac{1}{256} : \frac{1}{121} : \frac{1}{64} : \frac{1}{31{\cdot}36} : \frac{1}{16} : \frac{1}{7{\cdot}84}$$

The corresponding illuminations are therefore approximately in the ratio

$$1 : 2 : 4 : 8 : 16 : 32 : 64$$

Thus increasing the aperture by one stop, say from f/8 to f/5·6, halves the required exposure time.

Increasing the aperture not only lets more light in but also increases the lens aberrations. With commercially available cameras, aberrations are normally kept within satisfactory limits at the full aperture of the particular camera. With large apertures, the focusing also becomes much more critical. Only rays for a particular object distance are in focus at one time. Thus, in photographing a scene, only part of it will be sharply focused on the film and objects nearer or further than a particular distance become progressively more out of focus. However a certain amount of out of focus is not readily noticed in the final photographic image and can therefore be tolerated. The distance between near and far objects for which a sharp enough image is formed is called the *depth of field*. With large apertures, rays cut the axis more steeply, thus making the focusing more critical and hence the depth of field less. So a camera with, say, a lens of 45 mm focal length set to give correct focus for an object at 3 m will give a sharp focus for objects from about 2·5 m to 3·9 m away for an aperture setting of f/2·8. For an aperture of f/16, everything beyond about 1·4 m will be in focus. Thus a small aperture (large f-number) gives a large depth of field but involves a long exposure time with increased risk of blurring due to movement, whilst a large aperture gives a short depth of field with consequent critical focusing but a short exposure time.

Little has yet been said on the actual magnitude of the focal length of the camera lens but it should obviously be fairly short in order to achieve a compact camera unit. To avoid expensive large diameter lenses, the focal length should also be short so that low f-numbers can be easily obtained. It has been found in practice that a focal length slightly greater than the diagonal of the film negative being exposed gives a satisfactory compromise between sufficient illumination and the necessary freedom from aberration. Thus a 35 mm camera, that is, one using 35 mm wide strip film and exposing an area of 24 x 36 mm, should have a focal length a little greater than about 43 mm. In addition it may be required to change the focal length slightly to modify the camera for close-up use. This can be done by mounting a low-power positive meniscus lens in front of the camera lens, as in *Figure 2.4*. These meniscus lenses are available with powers of 1, 2, or 3 diopters, where the power of a lens measured in diopters is given by the reciprocal of the focal length as measured in air, expressed in metres. The supplementary lens should be as close as possible to the entrance

Table 2.1 TYPICAL OBJECT DISTANCES OBTAINABLE WITH CAMERA SUPPLEMENTARY LENSES

Supplementary lens power	*Camera focus set at* (m)	*Lens-to-object distance* (mm)
1 diopter	∞	1 000
	2	667
	1	500
2 diopter	∞	500
	2	400
	1	333
3 diopter	∞	333
	2	286
	1	250

pupil of the camera and has the effect of reducing the working distance (lens-to-object distance) with a consequent increase in size of the image on the film. By varying the camera focus a range of working distances can be obtained, as shown in *Table 2.1*; a complete table of distances is normally supplied with a supplementary lens by the manufacturers.

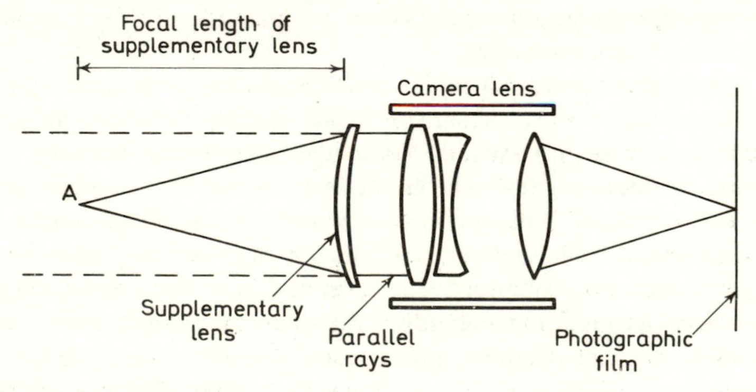

Figure 2.4. A camera focused for rays from an infinitely distant object is enabled to focus on a near object at A by the addition of a supplementary lens

However, it must be remembered that the depth of field becomes very small with shorter working distances. For example, with the working distance of 250 mm using the 3 diopter supplementary lens, the depth of field is about 30 mm with an aperture of f/16. Because of this small depth of field, coupled with the increase in aberrations owing to the simple supplementary lens, the aperture should be as small as possible compatible with a reasonable exposure time. Again, owing to the small depth of field, it is important that the object be at the correct distance from the lens. The distances given in *Table 2.1* are based on 'thin' lens calculations. Thus, with the camera focus set at infinity, the working distance is measured from the pole of the supplementary lens, as in *Figure 2.4*, but with the camera focus set at short distances, the working distance must be measured from the optical centre of the combined camera lens–supplementary lens system, which for most practical cases may be taken as the position of the variable aperture stop.

Rather than rely on measured working distances, it is far more satisfactory for scientific purposes to use a reflex type camera in which the image can be directly viewed, as in *Figure 2.5*. The image formed by an identical lens to the main camera lens is viewed on a ground glass screen, as in the two-lens reflex camera of *Figure 2.5 (a)*. In this, focusing adjustments cause both lenses to move together. It has the advantage that the object is continuously in view, even at the instant that the photograph is taken, but has the disadvantage that the view as seen in the viewfinder is from a slightly different position than that seen by the main lens, resulting in parallax difficulties when short working distances are involved. A better arrangement is to use a single-lens reflex camera, as in *Figure 2.5 (b)*. In this the object is viewed through the same lens as is used for photographing, with the mirror flicking out of the way at the instant the photograph is taken. *Figure 2.5 (b)* also shows an alternative viewing screen utilising a pentagonal prism. This has the advantage of showing the image the right way up and the right way round, whereas the simple ground glass screen gives lateral inversion which tends to make the following of moving objects difficult.

In these cameras the actual exposure time is controlled by a calibrated shutter which may be incorporated within the lens assembly close to the variable aperture or, alternatively, comprises a focal-plane shutter in the form of a roller blind arrangement placed immediately in front of the film. In the single-lens reflex camera, the action of pressing the exposure button first moves the front silvered mirror out of the way and then exposes the film. Most cameras incorporate a double-exposure prevention feature in which the action of moving on the film for the next exposure also cocks the exposure mechanism. Also generally incorporated are flash-gun contacts which can synchronise the

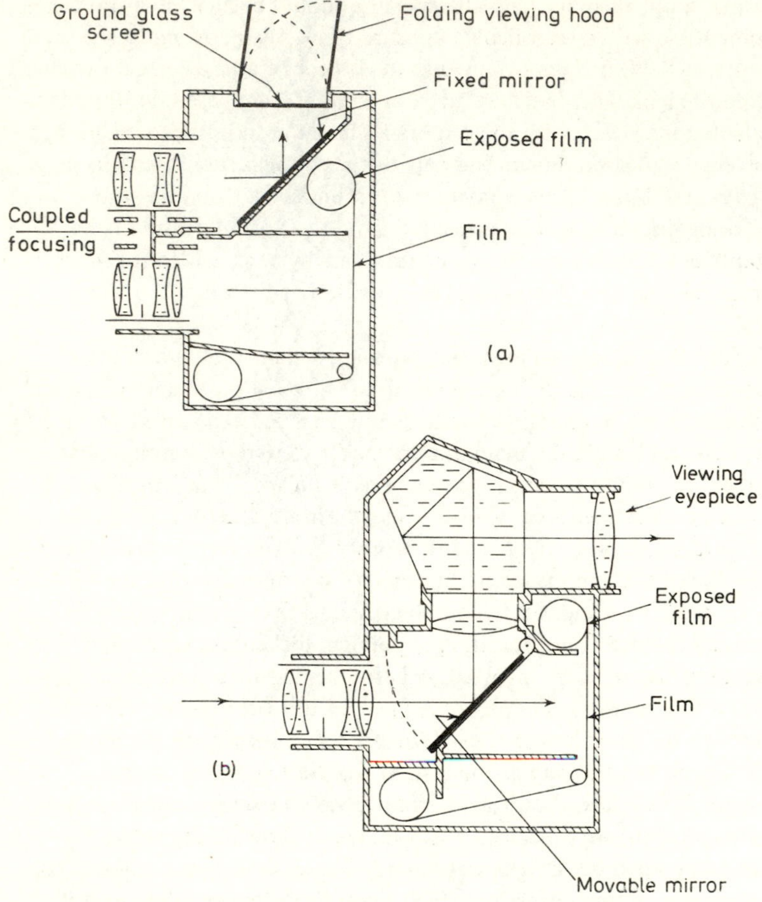

Figure 2.5. Arrangement of (a) twin-lens and (b) single-lens reflex cameras

firing of a flash bulb to the operation of the exposure button. Using a flash bulb, the shutter speed should be set to 1/30 s and the aperture set to an f-number determined by dividing the 'guide number' specified on the flash-bulb package by the camera-to-object distance.

2.2 *Exposure Meters*

Exposure times are best determined using an exposure meter. With this the scale of the meter is set to correspond to the speed of the film as recorded on the film package and then the light intensity is measured by the meter. This is recorded as a series of exposure times coupled with corresponding f-number apertures. It is then left to the user to choose the most suitable combination for his camera and the subject being photographed.

The barrier-layer type photo-electric exposure meter consists of a layer of selenium on an iron plate with the exposed face of the selenium covered by a thin gold vacuum-deposited layer, as in *Figure 2.6 (a)*.

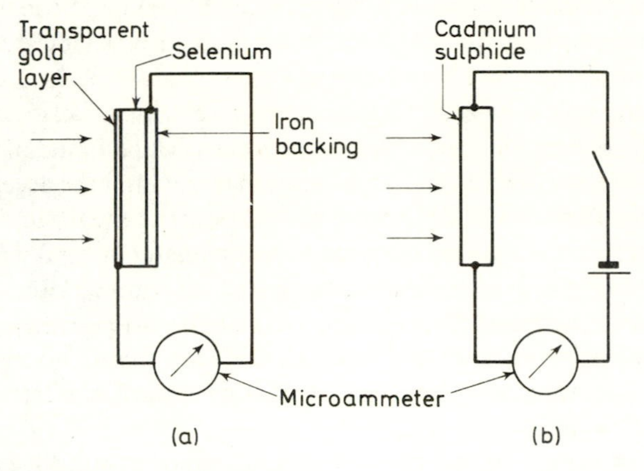

Figure 2.6. Arrangement of (a) selenium and (b) cadmium sulphide photo-electric exposure meters

This gold layer is sufficiently thin to be transparent but is still electrically conducting. Light falling on the gold side produces a current at the junction between the gold and the selenium semiconductor, which is measured by a microammeter connected between the gold and iron layers. The magnitude of the current produced is a measure of the intensity of the light falling on the cell. An alternative design of photo-electric exposure meter uses a plate of cadmium sulphide connected in series to a battery and microammeter [*Figure 2.6 (b)*]. In this type, light falling on the cadmium sulphide causes a change in resistance with a consequent change in current as recorded on the meter. The selenium type has the obvious advantage that no battery is required.

The exposure meter may be used to record either reflected or incident light. With the reflected light method, the exposure meter is pointed directly at the object to be photographed. The meter will then record an average value for the light being reflected. Different colours with widely differing reflectivities will be integrated to give a meter reading which assumes an average subject with the reflected light being evenly distributed. This reading is quite acceptable providing the subject is in fact an average one with an even distribution of reflectivities. If, however, the brightness range over the subject is very great with, say, only a small area of low reflectivity, then the reflected light reading will indicate an exposure that is satisfactory for most of the subject but will indicate too short an exposure for the dark region. It is, therefore, necessary to direct the exposure meter towards the brightest highlight region and the darkest shadow in turn, preferably taking the meter close to the subject to do this. If the range of exposures indicated is greater than about 120:1 if a black-and-white film is to be used, or about 8:1 for colour reversal film (transparencies), an unsatisfactory photograph will be obtained unless some means can be found of lightening the shadows. If the brightness range is within the acceptable limits of the film, then the exposure should be set to the mean of the two readings. Some cameras have the exposure meter integral with the camera so that the aperture and shutter speed are automatically controlled by the meter. This is quite adequate for average subjects with a low brightness range and in which there are approximately equal proportions of light and shade, but an override control is necessary for more versatile use.

The alternative method, which is generally recommended for use with colour reversal film, is to obtain incident-light readings. For this the light-sensitive part of the meter should be fitted with an incident-light attachment, which is simply a white translucent cover. The meter is then taken to the position of the subject to be photographed and the meter pointed at the camera. The white translucent cover is then illuminated to a level determined by the incident light and it is this white cover that is 'seen' by the exposure meter cell. The meter then records the exposure to be given to the subject as a whole and assumes that it is evenly illuminated by the same level of incident light. Areas that are away from the position of the exposure meter may well have a different level of illumination, possibly due to shading, and for these an incorrect exposure will be recorded. However, the exposure is correct for the particular area being monitored. For meters fitted with flat incident-light attachments, it is sometimes recommended that a reading also be obtained with the meter pointing towards the light source and the exposure based upon the mean of the two readings – one towards the source and one towards the camera. With domed covers this is not necessary, as the two readings are virtually the same.

In photography for scientific purposes, the cost of film material is usually negligible, unless vast numbers of photographs are involved, in comparison with the labour costs in producing the object to be photographed. Consequently it is common practice to determine what is thought to be the most suitable exposure conditions using an exposure meter and then taking extra photographs with slightly longer and slightly shorter exposures and under different lighting conditions in order to obtain the optimum. A photograph for scientific purposes is usually required to record a particular feature, sometimes to the photographic detriment of other features.

2.3 Photographic Film

The sensitive emulsion on a photographic film is not in fact an emulsion in the chemical sense but a suspension of silver bromide crystals in a gelatine base. The manufacturing process involves dissolving the gelatine in a solution of potassium bromide with some potassium iodide. Into this is stirred a solution containing silver nitrate, causing silver bromide particles to be precipitated. Depending on the concentrations of solutions, temperatures, and rates of mixing, the size distribution of the particles of silver bromide can be controlled, which in turn determines the speed and sensitivity of the film to light. Generally, the greater the grain size, the greater the speed and vice versa. Unfortunately, large grain size also means poor resolution which becomes more noticeable on enlargements, so that in practice it is necessary to choose a film that gives a satisfactory compromise between speed and resolution. To the emulsion mixture is added further gelatine, so that on cooling a firm jelly will be formed, as well as hardening agents, wetting agents, and sensitising dyes to extend the colour sensitivity. This sensitive material is then coated onto a suitable support material such as film, glass, or paper.

Depending on its required use, a modern film may be composed of a number of layers, as in *Figure 2.7*, which shows a section through a film suitable for cine use. The anti-static layer prevents the build up of electrical charge caused by rapid winding onto spools, which would otherwise cause dense markings on the film due to localised discharges. The anti-halation layer reduces back reflection of light into the emulsion, which would cause haloes and flare marks. The outer gelatine layer protects the sensitive layers from friction and abrasion marking.

On exposure to light and depending on its intensity, some of the silver bromide crystals are converted to silver with the liberation of bromine. The quantity of silver produced is, however, far smaller than that required to produce a visible image. Other grains exposed will also be modified, not in any visible manner but by being made more susceptible to chemical reduction. The photographic image at this stage is thus recorded as a *latent image*. The developing process involves the chemical reduction to metallic silver of those sensitive grains which have been rendered developable by the action of the light. The developer will

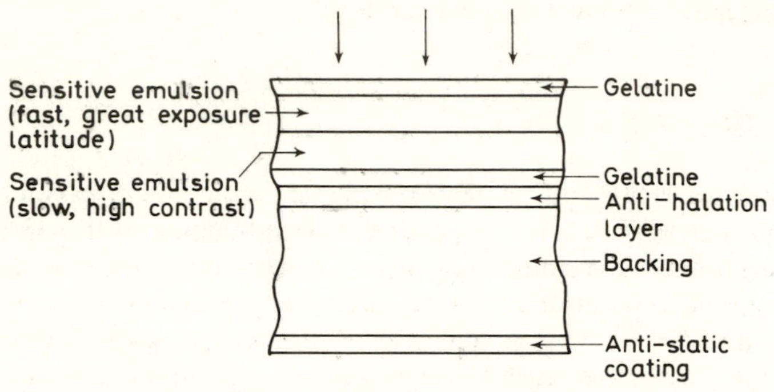

Figure 2.7. Section through a cine negative film, showing its possible complexity

in fact reduce all the grains to silver if given sufficient time but those grains which have been exposed to light will be reduced at a faster rate, depending on the amount of light exposure they have received. After sufficient development of the image has been obtained, the developing process is stopped by immersing the film in an acid-stop bath which may also contain a gelatine hardening agent, and the image is then fixed in a solution of acidified sodium thiosulphate in water. Further washing in running water is then necessary to remove all unwanted chemicals, after which the film must be allowed to dry in a dust-free atmosphere to allow the gelatine to harden. In practice, proprietary developing and fixing chemicals are used and the manufacturers' dilutions, temperatures, and times should be adhered to and related to the type of film used.

A number of factors may be used to measure the performance of a photographic film. The first of these is density, D, which is defined as the logarithm of the opacity, where opacity is the ratio of the incident to transmitted light intensity, that is,

$$D = \log_{10} \frac{I_0}{I} \tag{2.3}$$

By plotting the density against the logarithm of the exposure, the product of the exposure time in seconds and the intensity of illumination in lux, the characteristic curve of the particular film is obtained. It shows graphically the darkening of the film produced by exposure and development. In addition there are always some grains present in the emulsion which are spontaneously reducible and constitute *fog* in the film, which together with some light absorption by the gelatine and film base, constitute a density even without exposure to light. Idealised representations of characteristic curves for varying development times are shown in *Figure 2.8*. Characteristic curves are

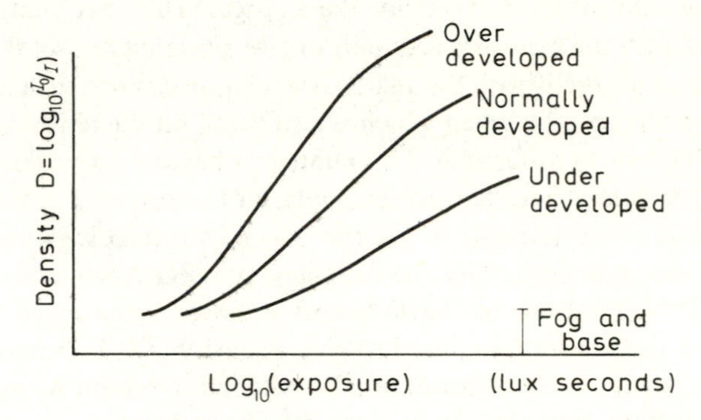

Figure 2.8. Idealised characteristic curves for different development times

also termed 'H and D curves' after Hurter and Driffield who first published such curves in 1890. For normal use, the films are used over the approximately straight region of maximum slope of their characteristic curve. The maximum slope of this curve, obtained under standardised exposure and development conditions, is termed the *gamma factor* of the film emulsion, and has values between approximately 0·5 and 5. In practice, the gamma factor represents the ratio of the contrast, or density differences, of the film negative to the contrast, or brightness, of the subject.

As well as speed, which is a measure of the rate at which the image is built up on exposure to light, colour sensitivity is an important factor in film performance. Two main types of black-and-white film are used. These are panchromatic, which has an approximately similar wavelength sensitivity to that of the eye, and orthochromatic, which is sensitive to green, blue, and violet, but not red light. Consequently, panchromatic film must be handled and developed in total darkness, whilst orthochromatic film can be handled in a dark room illuminated by a low-intensity red light. Orthochromatic film is especially suitable for recording from cathode-ray oscilloscope screens which usually fluoresce in blue or green.

Speed is measured on various arbitrary scales. The ASA scale, as devised by the American Standards Association, gives a number for the film speed which is based on an arithmetic scale. Thus a doubling or halving of a number indicates a doubling or halving respectively of the film speed. Most colour films have speeds in the range 16 to 160 ASA, whilst black-and-white films are made with a much greater range of speeds (up to about 1250 ASA), but special purpose films are obtainable with speeds well outside these ranges. The ASA number is concerned with the exposure required to give a minimum useful density difference. With the British Standards Association devised scale, a BS number of film speed is given which is also based on the requirement for a useful density difference. This number is based on a logarithmic scale (to the base 10) so that, for example, an increase of 3 in the BS number indicates a doubling of the film speed. A similar logarithmic scale has also been devised by the German Standards Association. This gives the DIN (Deutsche Industrie Norm) number for the speed and is a measure of the exposure required to give a density of 0·1 above fog. Another system uses the Weston number which is given on Weston exposure meters. However, on modern Weston meters this number is numerically the same as the ASA number. Generally only ASA and DIN

Table 2.2 COMPARISON OF ASA AND DIN FILM SPEEDS

ASA	*DIN*	*ASA*	*DIN*
16	13	160	23
20	14	250	25
32	16	320	26
40	17	500	28
64	19	650	29
80	20	1000	31
125	22	1250	32

numbers are now quoted for film speeds; *Table 2.2* gives a comparison of these speeds.

The production of a positive print on paper of a film negative follows a similar process to that used for originally obtaining the negative. An image of the negative is obtained on paper which is coated with a light-sensitive emulsion which is basically similar to that used on the film. The image may be obtained by contact printing, in which the paper is exposed to light while held in contact with the film negative, or by using an enlarger. With this the image is projected onto the sensitive paper and considerable enlargement may be obtained. After exposure the paper must be developed, fixed, and washed in the same manner as required for the film. Again the instructions as regards times and conditions as issued by the makers of the paper, developer, and fixer should be followed.

In colour photography the three primary colour components of a light source are individually recorded. If three light sources, a blue, a

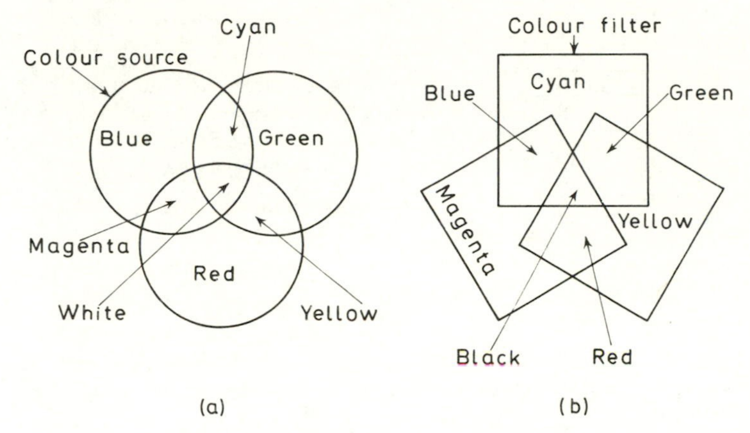

Figure 2.9. Results of (a) adding three primary-colour luminous sources, and (b) subtracting colours from white light by means of filters

green, and a red are projected onto a white screen so as to overlap, then as is well known white can be produced as in *Figure 2.9 (a)*, providing the intensities of the sources are suitably balanced. By adjusting the intensities, any colour of any required intensity may be produced. This is termed an *additive process*. If instead a white source is now used, with overlapping coloured filters placed to intercept the light falling on the white screen, a situation as in *Figure 2.9 (b)* may be obtained. It is seen that by, for example, using a yellow filter, the blue primary is removed from the white light to leave yellow, so that the yellow filter may be regarded as a 'minus blue' filter. The blue and yellow are *complementary* colours and it will be seen that any colour plus its complementary will produce white. Where the three filters overlap there is an absence of the three primary colours, leaving black. This is termed a *subtractive process* and, as before, by a suitable combination of the three subtractive primary filters, all colours and intensities may be obtained. It is this subtractive process that is the basis of colour photography.

Colour film is made up from three layers of sensitive emulsion on a suitable backing material such as film, glass, or paper, with each layer being sensitive to one of the primary colours, as in *Figure 2.10*. Ideally each layer should be sensitive to its own colour but in practice this ideal

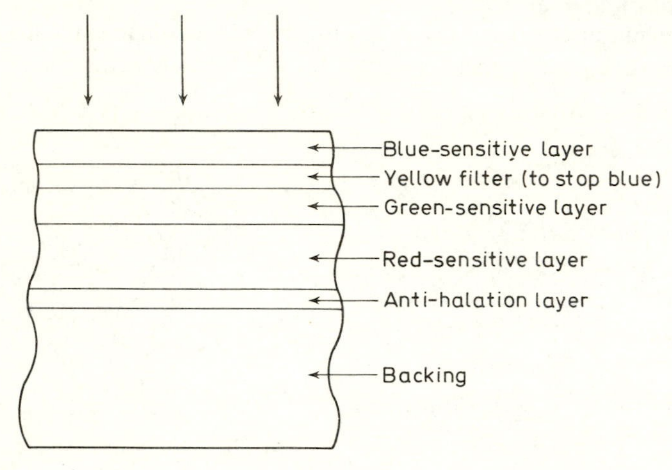

Figure 2.10. Section through a colour film to show the sequence of light-sensitive layers

is not strictly attained. The blue layer is the most selective and this is the reason for its coming first in the series. Because the green and red layers are also slightly sensitive to blue, a yellow filter layer is interspersed to remove the remaining blue light. This yellow coloration is removed during the developing process. Between the sensitive layers and the backing material is incorporated an anti-halation layer to reduce or prevent back reflection of light into the sensitive layers of emulsion. For printing-paper, since the requirements for exposing and processing are different from those for a film negative, the yellow filter and anti-halation layer may be omitted and the sensitive layers placed in a different order from that required for a film or glass plate base.

Colour balance can be achieved by using emulsions of various speeds so that, for example, in 'daylight-type' film the blue-sensitive layer may be made comparatively slow and the red-sensitive layer comparatively fast to compensate for the preponderance of blue over red in the composition of daylight. Conversely, for transparencies to be projected, a comparatively fast blue-sensitive layer and a slow red-sensitive layer are required to compensate for the excessive amount of red light emitted by tungsten-filament projector lamps.

The mechanism of image formation in the emulsion is the same as for black-and-white film. A black-and-white silver image is still produced; what is different with colour film emulsion is the presence of what is termed a colour coupler. This is an incomplete dye chemical which is different for each colour sensitive layer. The action of the colour developer is to react with each colour coupler, causing it to exhibit the colour relevant to each layer, whilst the density of the silver image determines the density of colour produced at each point within each layer. A further processing stage removes the silver image, leaving only the coloured image. The three layers of primary coloured images then give as a combination the correct colour and intensity of the final image. Depending on the type of film, a positive colour transparency may be obtained or a colour negative from which a colour positive on paper may be produced. A print made on paper does have the advantage that enlargement, as well as a number of copies, is possible, although the colour rendering is less brilliant than that possible on a positive transparency.

2.4 Polaroid–Land Process

The Polaroid–Land process first became generally available in 1948. It has the enormous advantage that the usual wet tank developing and fixing processes are eliminated, enabling a finished print to be obtained within seconds of the exposure. The camera used is basically of a conventional nature but with the addition of a film processing box on the back. Most modern Polaroid–Land cameras use film packs rather than rolls but the processing methods are similar. A film pack contains both negative film and positive paper. The negative is exposed in the usual way and then, by pulling a tab, the exposed negative is brought into face-to-face contact with the positive paper. A second tab is then pulled which draws both the negative and positive from the pack. As they emerge they pass between rollers which burst a pod of developing agent contained between the negative and positive and squeeze the agent uniformly over the interface. During the developing process a positive image is formed from the exposed film negative. The negative and positive have an opaque backing to protect them as they emerge from the film pack rollers. Developing should be allowed to continue for the time specified with the film pack and is about 10 s with a black-and-white film. For extra contrast or for lower temperatures, a slightly longer time is required. The two layers are then separated to obtain the positive print. To stabilise the print it must be coated by mounting between covers provided which, when wiped over, release a fluid which permanently fixes the print. It is not essential that the coating should be done immediately after developing although it should be done within a few hours. Care is required with the unused film packs as they should not be stored in a warm or humid atmosphere; if necessary they may be stored in a refrigerator. An advantage of this process is that it is possible to obtain very high-speed emulsions without the excessive grain size which limits the resolution of conventional high-speed film.

For conventional colour film the processing takes about $1\frac{1}{2}$ h to complete, but with the Polaroid–Land system a colour print is available in less than 1 min with the processing being done by the chemicals incorporated in the film pack as before. The chemical processes are, however, far more complex for colour processing than for black-and-white. After exposure the tab at the side is pulled to remove the negative–positive pair, the specified time allowed to elapse, and the pair separated to give the colour print. Unlike the black-and-white process, the colour print requires no further coating to stabilise it. The temperature of the film pack during the actual processing is of importance and it is necessary to adjust the processing time accordingly. Similarly the speed of the film varies with temperature and with large temperature variations it is necessary to adjust the exposure times.

For scientific use the advantage of having a print almost immediately available is obvious. Both copies and enlargements can be made and automatic cameras are available in which the period that the shutter is held open is controlled electronically, allowing satisfactory results to be obtained by persons with very little photographic experience or ability.

2.5 *Photocopying*

Documents may of course be copied using ordinary photographic methods with a camera and a 'close-up' lens but this process is obviously slow and costly. It does, however, have the advantage that the size of the copy can be altered. Thus microfilming is used for the recording and subsequent storage of very bulky material but, for general use, a quickly and simply obtained same-size copy of a document is required.

The actual fixing of the image on paper produced from the original document varies considerably, depending on the different manufacturers of equipment, but the printing can be done in one of two main

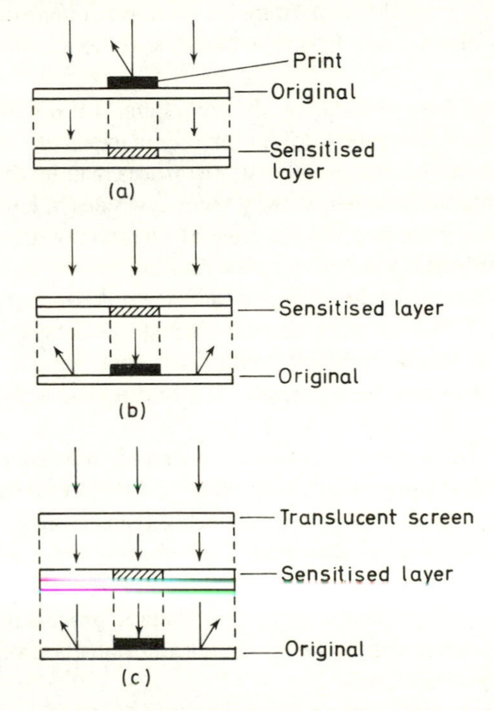

Figure 2.11. (a) Transmission, (b) reflex, and (c) indirect reflex printing arrangements for document copying

ways. If the document is printed on one side only, on reasonably translucent paper, then a transmission method may be used, as in *Figure 2.11 (a)*. Here light is passed through the document onto the copy paper which is in contact with it. Thus, wherever the copy paper is shielded from light by the dense print, its sensitised surface will be unaffected and on developing forms the image which will be the correct way around. More usually, however, the document has print on both sides and a reflex method must be used. In this the light passes through the sensitised copy paper onto the document. Light falling on the dark print will be absorbed whilst light falling on the unprinted areas will be reflected back onto the sensitised copy paper to form a laterally reversed copy, as in *Figure 2.11 (b)*. A readable copy may be made from this intermediate one either by it being put through the reflex printing cycle again or by being reversed and a copy made by transmission printing. Alternatively, using the arrangement shown in *Figure 2.11 (c)*, a readable copy may be obtained in one step by an indirect reflex method but usually with this some loss in definition occurs.

One of the earliest methods of photocopying, still widely used, is a reflex printing process using photographic type paper. By using silver type emulsions of reduced sensitivity, the process can be carried out in ordinary lighting but preferably away from a window. The intermediate copy or negative is fed into the machine by electrically driven rollers which pass it through developing and stabilising trays of solution. While still wet this negative can be used to produce a positive copy by reflex printing again. This copy, after passing through the developing and stabilising trays, must then be allowed to dry. Any number of copies may be produced from the negative and should remain stable for several years.

A further development is the diffusion transfer process which eliminates the development and stabilising of the intermediate negative. Using a reflex method as before, the sensitised paper and original document are exposed to light to obtain a negative latent image. This negative is then placed in contact with a sheet of positive transfer paper and fed by rollers into the machine for automatic processing, where the negative is developed and the image chemically transferred onto the positive and then stabilised.

An ideal method of copying drawings is the dyeline process. The original to be copied should be on a translucent medium and with a good line contrast. Diazonium-coated copy paper is used which is only sensitive to ultraviolet light. A straightforward transmission copy is made which has then to be developed either by passing through a suitable liquid or by exposure to ammonia vapour, depending on the type of paper used. The developing makes the residual diazonium coating visible wherever it has been protected by the print from being bleached out.

More modern processes have concentrated on simplicity of use and if possible the avoidance of wet bath developing and fixing. With the

'Thermo-fax' process, for example, the sensitised paper only reacts to infrared light and has the advantage that it can be used in brightly lit surroundings. The sensitised paper and document to be copied are placed in contact and exposed to an infrared source. Any print, which should be dense and from a carbon or metal based ink, will absorb the infrared rays and the localised heating will activate the sensitised paper to form a darkened image. Where there is no absorption there will be no localised heating and no colour change.

Finally, mention may be made of the xerographic method of printing, using electrostatic effects. In this process a plate coated with selenium is first given a positive electrostatic charge; onto this plate is then projected an image of the document to be copied, using a lens system. Areas on the selenium plate that are exposed to light will then become conducting and lose their charge whilst dark areas, corresponding to the print, will retain their charge. Negatively charged powdered ink is then sprayed onto the plate where it will adhere to the positively charged print image. Now, by bringing up a positively charged sheet of ordinary paper, ink will be transferred to the paper where it may be made permanent by heating. By controlling the amount of ink used about 15 copies may be made from the one plate. For re-use, the plate is simply wiped clean and recharged. Commercial versions of this process use a drum rather than a plate and the whole process is made quite automatic.

OPTICAL INSTRUMENTS

2.6 Telescopes

Essentially a telescope consists of two lenses mounted coaxially: a long-focus lens, or objective, and a short-focus lens, or eyepiece. In practice these are both compound lenses or systems. In the *astronomical telescope* arrangement shown in *Figure 2.12 (a)*, parallel rays incident from a distant object at an angle θ to the optic axis will be brought to a focus at A′, where OA is equal to the focal length f_O of the objective. It is usual to position a stop at A to obtain a clean edge to the field. The positive eyepiece lens, E, is situated at a distance such that AE is equal to f_E, its focal length. Thus the rays are deviated by the eyepiece to form parallel rays at an angle θ' to the axis, that is, the angle subtended at the eye by a distant object has been effectively increased, causing the object to appear larger. It is immaterial whether the angles are considered to be subtended at the eye or the telescope objective, since the length of the telescope is negligible compared with the object and

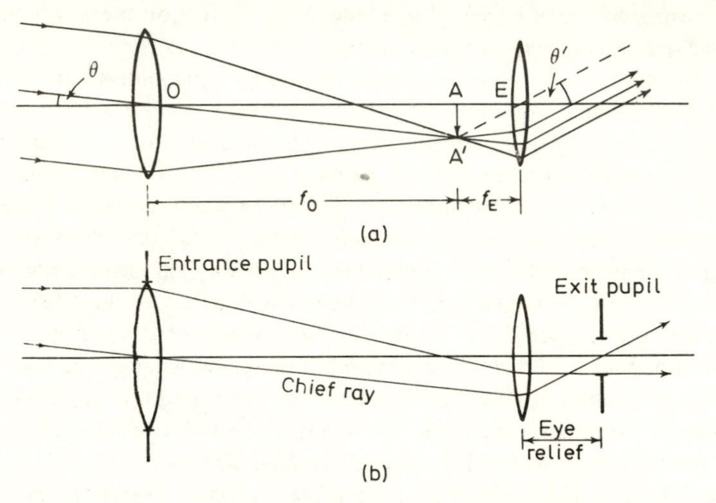

Figure 2.12. Ray paths through an astronomical telescope

image distances. Since these distances may be considered infinite, they cannot be used to calculate the magnification. The only magnification with meaning involves angles. Thus the angular magnification may be defined as

$$\frac{\theta'}{\theta} = \frac{\tan\theta'}{\tan\theta} = \frac{AA'/f_E}{AA'/f_O} = \frac{f_O}{f_E} \tag{2.4}$$

where θ and θ' are small. It is also common practice to call this angular magnification 'magnifying power' or simply 'magnification'. Since the object and image rays are inclined in opposite senses to the axis, the image is inverted and this is the reason that the telescope is termed 'astronomical', since for astronomical use an inverted image is of no consequence.

If it is assumed that the edge of the objective lens limits the cone of rays entering the telescope, then the objective may be regarded as the aperture stop. In this telescope it is also the entrance pupil and the exit pupil is its image formed by all the lenses to the right of the aperture stop, which in this case is only the eyepiece. A chief ray and a ray parallel to the axis, which together define the diameter and position of the exit pupil, are shown in *Figure 2.12 (b)*. A chief ray is simply a ray that passes, or appears to pass, through the centre of the entrance pupil and through the centres of both the aperture stop and exit pupil. By a comparison of similar triangles, the magnification

$$\frac{f_O}{f_E} = \frac{E_n}{E_x} \tag{2.5}$$

where E_n and E_x are the diameters, respectively, of the entrance and exit pupils and where E_n is the diameter of the clear aperture of the objective. This gives a quick way of establishing the magnification of the telescope. By focusing the telescope for infinity and pointing it at the sky, the exit pupil may be located by moving a screen back and forth behind the eyepeice until a position is found where the emergent beam is at its smallest and most sharply defined. This then gives the diameter and position of the exit pupil and hence the magnification may be calculated. Caution should be exercised, however, as on no account should a telescope be used to look directly at the sun, otherwise serious damage may be done to the eye.

The exit pupil is also called the *eye circle* or *Ramsden circle* and is the position at which the eye pupil should be placed for correct viewing. If the diameter of the exit pupil is greater than that of the eye pupil, not all of the light will enter the eye. The maximum illumination of the retinal image occurs when the pupils are the same size but, in practice, it is found to be better to have the exit pupil slightly smaller than the eye pupil for ease and comfort of viewing. For daylight viewing, the overall magnification and the diameter of the objective are arranged to give an exit pupil of 1·5–2 mm diameter. The distance between the eye and the eyepiece – that is, the distance of the exit pupil from the eyepiece – is the *eye relief*. The optimum eye relief depends on the use of the telescope. Where comfort is the only consideration, a distance of about 8 mm is usually satisfactory but if, for example, the telescope is to be used for gun sighting, the eye relief is greater to allow for recoil. The actual amount of eye relief is simply calculated. The exit pupil is the image of the entrance pupil as formed by the eyepiece. Therefore, with respect to the eyepiece, the object distance is $u = -(f_O + f_E)$, as in *Figure 2.12 (b)*, and the eye relief is the image distance v. Hence, applying the thin-lens formula $1/v - 1/u = 1/f$,

$$\text{eye relief} = \frac{f_E(f_O + f_E)}{f_O} \tag{2.6}$$

To increase the magnification of a telescope, it is necessary to have an eyepiece of short focal length or an objective of long focal length. An eyepiece of very short focal length presents difficulties – for example, the

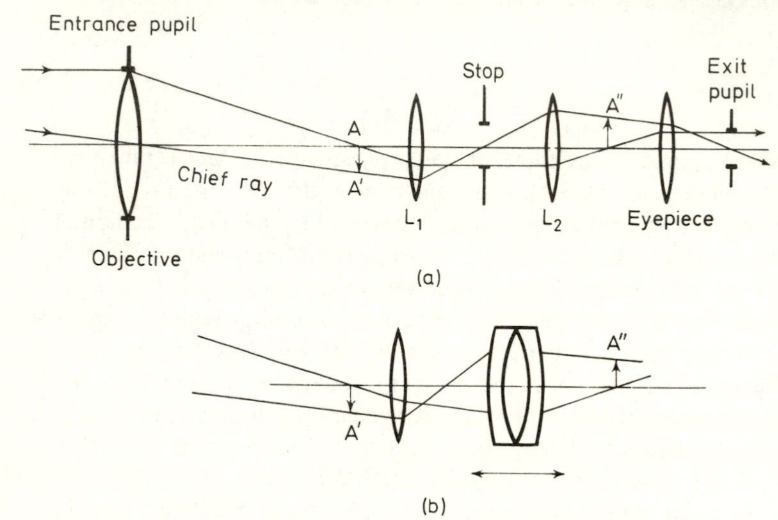

Figure 2.13. Ray paths through a terrestrial telescope

eye relief distance becomes short and the positioning of the eye pupil becomes critical – whereas an objective of long focal length involves small angles of refraction and consequently a high degree of aberration correction can be achieved. The nature of the possible aberrations has already been discussed in Section 2.1. By using a cemented doublet, the objective lens can be corrected for spherical and chromatic aberration but not usually for coma, although this can be quite small. The elimination of coma is not, however, always essential since for many telescopes only good central definition is important. For large aperture objectives, two slightly separated lenses or a combination of cemented and separated lenses are often used to give greater freedom in design.

A *terrestrial telescope* is an astronomical telescope with an additional inverting element, which can be a system of lenses or prisms, to give an erect image. In *Figure 2.13 (a)* two similar lenses L_1 and L_2

have been added to the basic telescope of *Figure 2.12 (b)* and are used to invert the image formed at A′, re-forming it at A″. This real image is then magnified in the usual way by the eyepiece. The erecting system and the eyepiece are then moved as one unit for focusing. Although single lenses are drawn for clarity, the objective is normally at least a doublet and the eyepiece a two-component combination.

By adjusting the separation of the erecting lenses, the magnification may be changed. The actual erecting lens is usually a cemented triplet as in *Figure 2.13 (b)*. If the erecting lens is moved towards the first image at A′, it produces a magnified image at A″, now further to the right. The eyepiece is then moved to focus on the image at the new position of A″. For objects not at infinity, the objective forms the image A′ at a different position. For focusing, the complete erecting system and eyepiece have to be moved as a unit.

By mounting the triplet erecting lens and eyepiece in telescopic tubes connected by a suitable scroll thread, the magnification of the telescope may be changed whilst still preserving the focus. The variable magnification with such telescopes is usually from about 5 to 20 times. A single corrected lens may also be used as an erector, in place of the two-lens systems so far discussed, but this is more suitable as a fixed-magnification erector as the lens can only be at its best aberration correction at a particular magnification.

The magnification of such telescopes is a combination of that of the basic astronomical telescope and that of the erector system. The magnification of the complete telescope is thus $M f_O / f_E$, where M is the magnification produced by the erector. To obtain a high magnification with this type of telescope, an objective lens of long focal length is required and this, when coupled to the erector system, yields a somewhat unwieldy instrument. A considerable saving in length can, however, be achieved by using a prism erecting system as is used in binoculars.

2.7 *Binoculars*

Figure 2.14 shows two of a number of alternative arrangements of prisms used to reverse a pencil of rays both left to right and from top to bottom. The insertion of such an arrangement between the objective and eyepiece of an astronomical telescope converts it for terrestrial use and considerably shortens its length. Two such identical telescope arrangements constitute a pair of binoculars. It is important to ensure that the emergent ray pencil is parallel to the incident rays, otherwise it is not possible to superimpose the two images from the separate telescopes without considerable eye strain. Generally, this means that the 90° prism angles should be accurate to ± 3′ and the 90° edge must be parallel to the hypotenuse face – that is, there must be no pyramidal error. As well as accurately made prisms, good aberration correction is required and therefore the objective is normally a doublet or triplet and

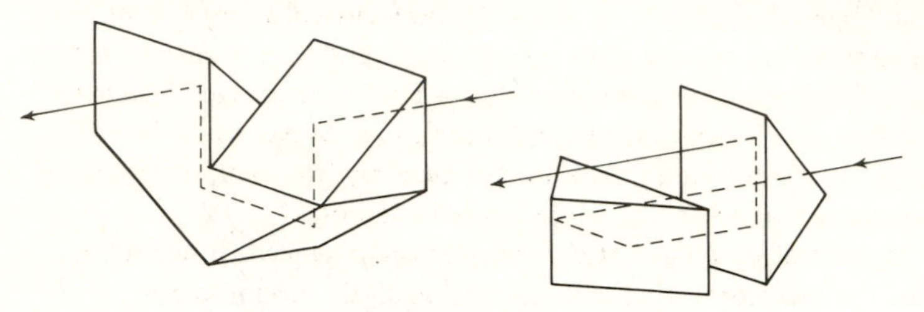

Figure 2.14. Prism erecting systems

the eyepiece a combination involving separated components. Since the path through the glass prisms may amount to several centimetres, small amounts of spherical and chromatic aberration, astigmatism, and coma are introduced. Chromatic aberration is corrected in the eyepiece and the others are dealt with by leaving corresponding amounts of residual aberrations of opposite sign in the objective. Both telescopes are focused together by a single central control. In addition one or both of the eyepieces may be focused independently over a calibrated range of about −5 to +5 diopters to allow for differences in the observer's eyes. A diopter is defined as the reciprocal of the focal length, where the focal length is measured in metres.

Because binoculars are usually held in the hand, they should be light in weight and compact. The 'folded' design is useful from this point of view and also brings their centre of gravity within the user's hands. It is difficult to hold binoculars with more than x 10 magnification sufficiently steady without a stand and so most hand-held instruments have a magnification between x 5 and x 8.

Binoculars should also have a large field of view and good light-gathering power. The field is controlled by the eyepiece aperture. For example, a 5° object field corresponds to a 5 x 10 = 50° image field for a x 10 magnification, which is quite large enough for normal requirements. The light-gathering power depends on the clear aperture of the objective. Providing the exit pupil almost fills the eye pupil and the glass surfaces are 'bloomed' to reduce the amount of light lost by reflection, the image should not appear noticeably darker than the object viewed direct.

It is usual to describe a pair of binoculars as, for example, 8 x 30. This means that they have a magnification of 8 times and an objective with a clear aperture of 30 mm. This gives an exit pupil of 3·75 mm. In twilight, the pupil of the eye has a diameter of about 7 mm and therefore 'night glasses' are scaled up in size to give an exit pupil of approximately the same size as the eye pupil. These binoculars can be used in daylight but are rather cumbersome.

The two halves of a pair of binoculars are hinged to adjust the interocular distance to suit the observer. Owing to the arrangement of the erecting prisms, the separation of the objectives is about twice this distance. This means that the radius of stereoscopic vision is double that

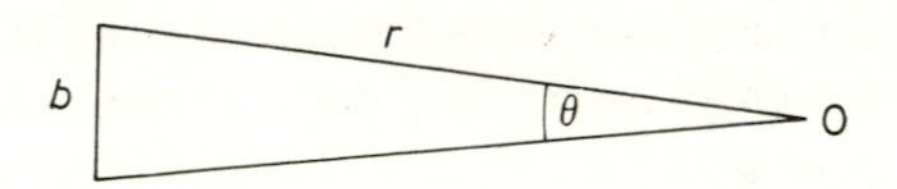

Figure 2.15. A distant object at O as viewed by a person whose interocular distance is b

due to the magnification alone. In *Figure 2.15*, O is a distant object viewed by a person whose interocular distance is b and $b \ll r$. Then

$$r = \frac{b}{\theta}$$

Stereoscopic vision just ceases when r is large enough for θ to equal α, the limiting angle of resolution of the eye. Thus stereoscopic vision just ceases when $r = r'$, such that

$$r' = \frac{b}{\alpha}$$

The *radius of stereoscopic vision*, r', is approximately 200 m with an interocular distance, b, of about 60 mm and a limiting angle of resolution $\alpha = 1' = 0{\cdot}00029$ rad. This means that points which are more than about 200 m from the eyes cannot be distinguished from points which are an infinite distance away, except by comparison of their size with known objects at known distances. With binoculars of magnification M, the limiting angle of resolution α is reduced to α/M and the interocular distance b is increased to b', the separation of the objectives. Thus the stereoscopic radius of the binoculars becomes S, where

$$S = \frac{b'}{\alpha/M} = \frac{Mb'}{\alpha}$$

$$= r' \frac{Mb'}{b} \qquad (2.7)$$

For binoculars giving a magnification of x10 and with the separation b' of the objectives twice that of the interocular distance b, the stereoscopic radius is increased 20 times over that of the eye alone, that is, from about 200 m to about 4 km (approx. $2\frac{1}{2}$ miles).

2.8 *Theodolites*

Telescopes with special properties are required for sighting and the accurate determination of angles. They must be able to stand up to use

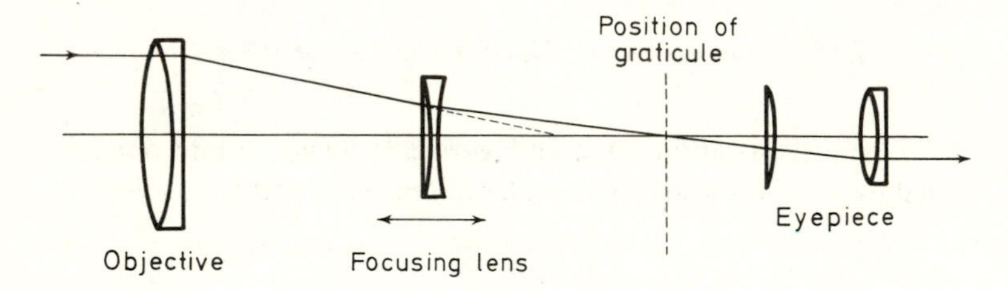

Figure 2.16. Internal focusing arrangement for a telescope

in wet weather and, since they are normally mounted on tripods, they must be compact and remain balanced as the focus is changed. These initial requirements can be met by an internal focusing arrangement, as in *Figure 2.16.* Since for measurement purposes it is of little importance if the image is inverted, an astronomical telescope with its smaller number of lenses can be used as the basic instrument. Instead of altering the distance between the objective and the eyepiece, the instrument is focused by adjusting the position of a negative lens between the objective and the eyepiece. This has the advantages that the centre of gravity changes only slightly and that the objective and eyepiece can be sealed into the telescope to keep out the weather. The internal focusing lens inside the main telescope barrel moves coaxially as its position is adjusted by means of a rack and pinion. The pinion shaft is fitted with a stuffing-box where it projects through the main telescope tube, again to keep out the weather.

Another advantage of the negative focusing lens is that it gives a compact telescope as in effect it increases the focal length of the objective, thus giving high magnification without the disadvantage of a long telescope tube or an eyepiece of inconveniently short focal length. Since high definition is required with such a telescope, the complete instrument should be well corrected for spherical and chromatic aberration and coma. Some astigmatism is, however, allowable at the edge of the field as the edge is only used as a finder.

For sighting purposes it is necessary to have some form of cross-line or graticule. This may be either ruled or formed by a photographic process onto a glass plate and located at the focal plane of the eyepiece.

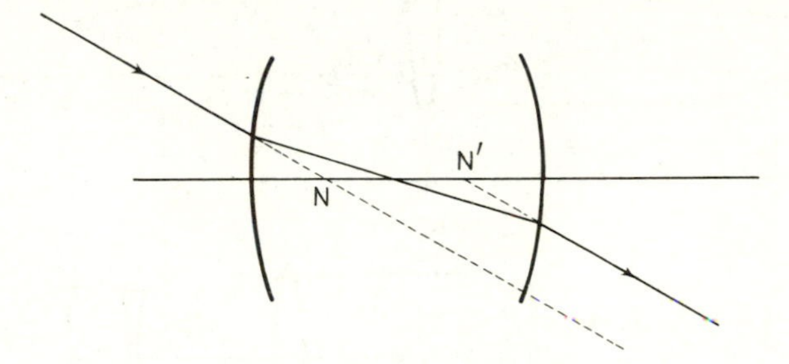

Figure 2.17. The nodal points of a thick lens

There are usually two parallel lines on the graticule which subtend an angle, usually 1 in 100, at the second nodal point of the objective. The positions of the nodal points of a thick lens are illustrated in *Figure 2.17*, where the incident and emergent rays are parallel and where N and N′ are termed respectively the first and second nodal points. The angle subtended by these two lines on the graticule is called the *stadia interval* and is equivalent to 34′ 22·6″ (2062·6″). To measure a distance the telescope is focused on a distant graduated staff; the length of staff seen between the two stadia lines is then approximately equal to 1/100 of the distance of the staff from the telescope. If the staff-to-telescope distance is changed, the telescope must be refocused. This means that the distance between the objective and its near image plane is changed and the stadia interval no longer subtends the correct angle at the telescope. However, *Figure 2.18* shows that, for the external focusing telescope – that is, one in which the eyepiece and the graticule are moved together for focusing – there is a point in front of the objective to which distances may be measured and agree correctly with the measurement using the stadia marks. The distance of this point from the mounting axis of the telescope is termed the 'constant'. If ab is the stadia interval corresponding to angle θ, then as the telescope is refocused the image is formed at some other plane, say, a′b′, but as is seen from the figure the same stadia marks correspond to the same angle θ no matter in which plane the image is formed. Thus all distances

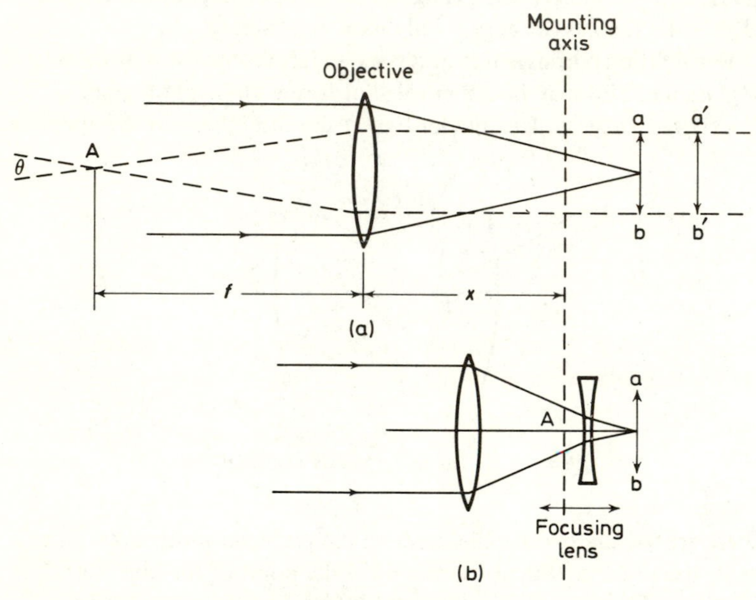

Figure 2.18. The position of the anallatic point A shown with reference to (a) an externally focusing and (b) an internally focusing telescope

must be referred to the point A, termed the *anallatic point*, with the distance $(f+x)$ being added to give the true distance of the staff from the mounting axis of the telescope. Providing the instrument is focused by adjusting the position of the eyepiece and graticule and not by moving the objective, this distance $(f+x)$ is constant.

However, with an internally focusing telescope it is possible to choose the telescope length and the powers of the objective and negative focusing lens such that compensating effects are introduced. These keep the angle subtended by the staff at the centre of the instrument, the mounting axis, the same as the angle subtended by the stadia marks, irrespective of the distance of the staff. That is, the anallatic point A now lies on the central mounting axis of the instrument, as in *Figure 2.18*, which also illustrates the comparative shortening in length of an internal focusing telescope over an external one of the same power. In practice, the compensation is not exactly correct but the error introduced is generally less than 1 in 1000 so that the 'constant' may normally be disregarded. Thus, if the staff intercept is 2·50 m, as measured by a 1-in-100 stadia interval, the distance of the staff from the centre of the theodolite neglecting the 'constant' is 250 ± 0·25 m.

Besides measuring distances by means of the stadia interval, theodolites are used for measuring the angles subtended by two distant objects by sighting on the objects in turn and measuring the angle subtended on a graduated circle. This requires the sighting of the telescope with the maximum accuracy possible. However, owing to diffraction (*see also* Section 4.2) the minimum angle which can be resolved by a telescope is equal to $1{\cdot}22\ \lambda/a$, where λ is the wavelength of the light and a is the diameter of the circular aperture which limits the beam, in this case the clear aperture of the objective. With an objective aperture of, say, 40 mm and with light of a mean wavelength of 6×10^{-7} m, the minimum angle of resolution with this telescope will be $1{\cdot}83 \times 10^{-5}$ rad (3·8″). Since the unaided eye can just resolve two points separated by about 1′, the eyepiece must have a magnification of at least ×16, so that the eye can see all that is resolved by the objective. A higher magnification than this does not provide any more information and gives a dimmer image.

The resolution of the eye is at its maximum when the pupil diameter is about 2 mm. This means that a telescope with a 40 mm diameter objective – that is, one with a 40 mm diameter entrance pupil – should have an overall magnification of ×20 to give a 2 mm exit pupil and hence fill the eye pupil. Together with the magnification required by the eyepiece, this is a satisfactory overall magnifying power, since it makes full use of the resolutions of both the objective and the eye and gives an eyepiece with a convenient focal length. Thus there is an optimum magnification depending on the clear diameter of the objective lens.

2.9 Eyepieces

Many eyepieces have been designed to serve many different specialised purposes but for most their use is strictly limited. However, two designs, the Huygenian and the Ramsden, have come into popular use and serve for the majority of optical instruments.

The purpose of an eyepiece is to magnify the image produced by an objective without causing a deterioration in the quality of the image and, if necessary, to correct any residual aberrations. Thus, for example, matched pairs of eyepieces supplied with binocular microscopes should not be separated and intermixed with other eyepieces nor should a compensating eyepiece supplied with a particular microscope objective be used with other objectives. An eyepiece should also give an eye relief distance – that is, the distance between the exit pupil, usually the last lens of the system, and the observer's eye – of about 8 mm for normal usage. If the eye relief distance is less than this, eyelashes get in the way, and if it is more there may be difficulty in positioning the eye pupil correctly, which would cause cut-off of part of the angular field. The image field produced by the eyepiece should also be flat to prevent undue eye strain and so that, for example, a graticule scale is sharply in focus over its whole length. It is normal to attempt to satisfy these requirements with a two-lens system. Eyepieces are commonly made with nominal magnifying powers of × 5, × 10, × 15, and occasionally higher powers for astronomical telescopes.

The *Huygenian eyepiece* consists of two plano-convex lenses made from glass of the same refractive index and mounted with the plane surfaces towards the eye, as in *Figure 2.19 (a)*. The ratio of the focal lengths of the field lens to eye lens f_F:f_E varies from 3:1 to 2:1 depending on the manufacturer, with the lenses being separated by a distance equal to half the sum of the individual focal lengths. It may also be shown that the spherical aberration will be a minimum when the separation is equal to $(f_F - f_E)$, although this condition is not usually satisfied except when the focal lengths are in the ratio 3:1. By separating the lenses by a distance equal to half the sum of the individual focal lengths, that is, by $\frac{1}{2}(f_F + f_E)$, it may be shown that the system can be made substantially free of chromatic error. If, however, the eyepiece is used with a graticule, this chromatic correction will be upset. To be in focus, the graticule has to be placed between the lenses in the plane of the field stop so that it is only viewed by the eye lens and consequently shows some coloration. One objection to this type of eyepiece, apart from its use with graticules, is that the eye relief is rather short for comfortable viewing. There is also some spherical aberration and astigmatism, and rather pronounced curvature of field.

The *Ramsden eyepiece* also consists of two plano-convex lenses but these now have the same focal length and their convex faces are towards each other, as in *Figure 2.19 (b)*. To satisfy the condition for achromatism, the lenses should be separated by a distance equal to half the sum of their focal lengths which, in this case, is equal to either of their focal lengths. This has the disadvantages that any dust on the surface of the field lens is sharply focused by the eye lens and that the eye relief is very small when used with a telescope. To overcome these

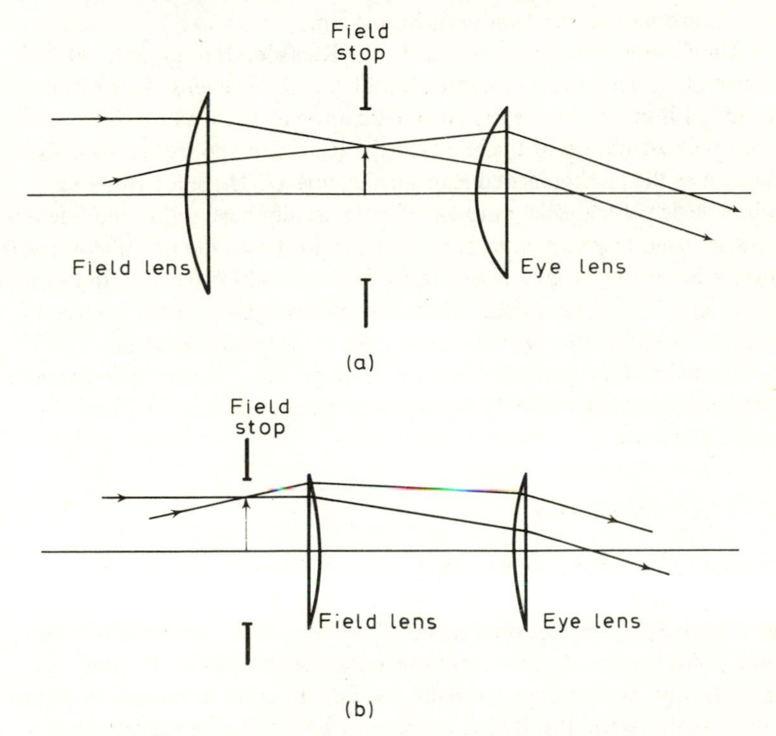

Figure 2.19. Ray paths through (a) a Huygenian and (b) a Ramsden eyepiece

difficulties the separation is usually made two thirds of either of the focal lengths and this, in fact, introduces only a small amount of chromatic error. The image produced by the objective, which now becomes the object for the eyepiece, is arranged to be in the plane through the first focal point of the eyepiece, which is also the position of the field stop. As the object is outside the system, a Ramsden eyepiece is used for all measuring purposes, since a graticule can be located in the plane of the field stop and magnified with the object without any additional chromatic or other aberration. Since the object is outside the system, this type of eyepiece is sometimes termed positive, whereas the Huygenian is termed negative.

The Ramsden and the Huygenian eyepieces have approximately the same angular field of view which is about 40° to 50° for a magnifying power of x10. The Ramsden, however, has somewhat less spherical aberration and distortion but, owing to the incorrect separation, has more chromatic aberration in its basic form.

The *Kellner eyepiece* is similar to the Ramsden but has a double-convex field lens and a cemented doublet as the eye lens. This form of eyepiece is corrected for the chromatic error introduced by the incorrect separation of the lenses in the Ramsden and is therefore also known as the 'achromatised Ramsden eyepiece'. There are other eyepiece designs for special purposes. These usually have rather more lenses and are used to give greater angular fields, in the region of 70° for a x10 eyepiece, or to give greater eye relief distances which may be up to the focal length of the eyepiece. These long eye relief eyepieces are usually meant for use in gun sights or to be used by spectacle wearers.

To make all eyepieces readily interchangeable, their outside diameter has been standardised to 31·8 mm for telescope use and 23·3 mm for microscope use.

2.10 Autocollimators

An autocollimator consists simply of a well-corrected lens with a pinhole aperture accurately positioned at a principle focus of the lens, as in *Figure 2.20*. On illuminating this pinhole, which then approximates to a point source, the lens forms a parallel beam of light. If these parallel rays are incident normally onto a reflecting surface, they return along their own paths. By means of a suitable half-silvered reflector, these rays are now reflected to one side of the collimator axis to form an image of the pinhole source, which in turn is viewed by a Ramsden eyepiece. The half-silvered reflector is in the form of a cube cut diagonally, polished, half-silvered, and cemented together again. This has an advantage over a plain glass plate half-silvered on one face since there is no ghost reflection from the other face.

The autocollimator is particularly useful for the measurement of small angles. If the reflecting surface used to return the parallel beam is tilted, the position of the image of the pinhole as seen through the eyepiece shifts to one side of its original position. Suppose, for example, that the reflecting surface is tilted through an angle of 1′, then the reflected ray is shifted through 2′. Using a collimating lens of, say, 0·5 m focal length, the image is shifted 500 tan 2′ = 500 x 0·0006 = 0·30 mm. This distance is readily measured by a magnifying eyepiece

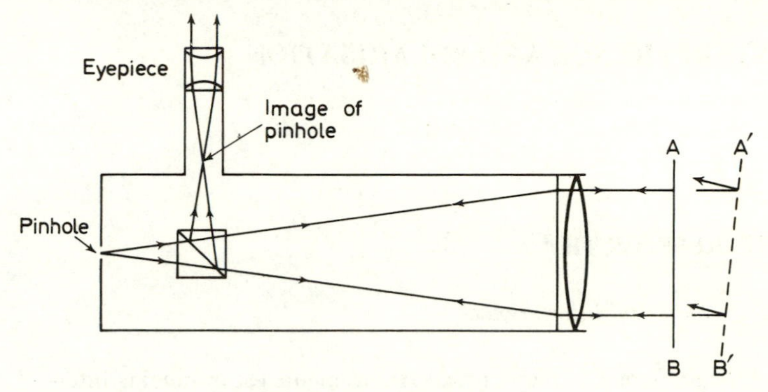

Figure 2.20. Arrangement of an autocollimator for the measurement of the small angle between the reflecting surfaces AB and A′B′

with a calibrated graticule. If, for example, the diameter of the pinhole is 0·20 mm, this is the size of the image. Thus, for a 1′ tilt, the image is shifted by $1\frac{1}{2}$ times its own diameter so that the graticule can be calibrated.

Three

INTERFERENCE AND POLARISATION

INTERFEROMETRY

3.1 Thin-film Interference

Although there are many optical arrangements for producing interference, it is the interference produced by multiple reflections within

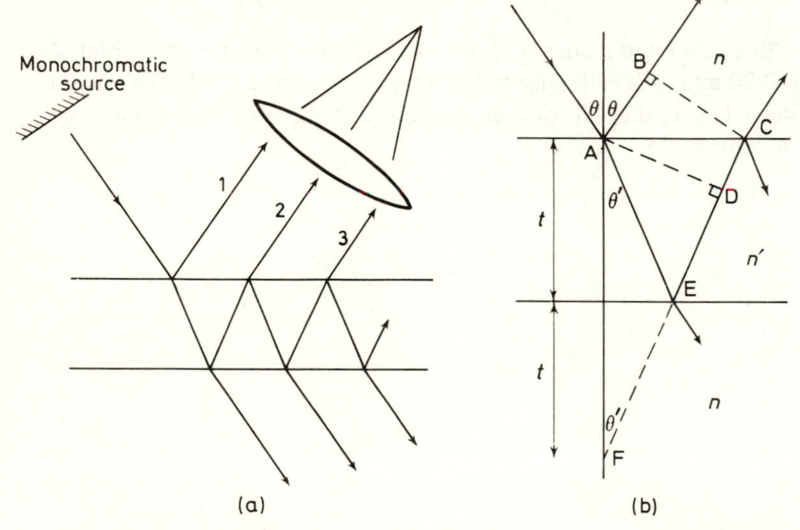

Figure 3.1. Multiple reflections occurring at a thin film of refracture index n′

thin films that is of the most practical use. This occurs when light from a monochromatic source falls, for example, on a thin parallel sided film so that some of the light is refracted into the film and some is reflected, as in *Figure 3.1*. The refracted ray is again split into a reflected and a refracted part at the lower surface, the reflected part being again divided at the upper surface. This repeated division at the upper and lower surfaces continues until the remaining intensity falls to zero. Thus a path difference is introduced between adjacent parallel rays, from either the upper or the lower surface.

From *Figure 3.1 (b)* it is seen that this optical path difference is equal to the distance AEC in the film of refractive index n', minus the distance AB in the medium of refractive index n. From the wave front BC onwards, no further path differences, and hence phase differences, will be introduced between the two rays. Hence the path difference between these two rays is

$$\begin{aligned} \text{path difference} &= n'(\text{AEC}) - n(\text{AB}) \\ &= n'(\text{FD} + \text{DC}) - n(\text{AB}) \\ &= n'(\text{FD}) \\ &= 2n't \cos\theta' \end{aligned} \tag{3.1}$$

since optical paths must be the same for any ray drawn between corresponding wave fronts, in this case AD and BC, so that $n'(\text{DC}) = n(\text{AB})$.

Whenever this path difference is equivalent to a whole number of wavelengths, $m\lambda$, a summation of amplitudes would be expected to lead to a maximum of intensity, that is, constructive interference would be expected. However, a phase change of π occurs with a ray that is reflected at a boundary with a higher refractive index medium. This phase change, it should be emphasised, only occurs when light strikes a boundary from the side of lower refractive index and not vice versa. For the film considered in *Figure 3.1*, it is ray 1 alone that undergoes a phase change of π. Thus

$$2n't\cos\theta' = m\lambda \tag{3.2}$$

where $m = 1, 2, 3, \ldots$, is in fact the condition for destructive interference between rays 1 and 2 and a minimum of intensity will occur for each particular value of m. Similarly, for constructive interference between these two rays producing a reflected maximum of intensity

$$2n't\cos\theta' = (m + \tfrac{1}{2})\lambda \tag{3.3}$$

However, rays 2, 3, 4, . . ., all approach reflecting boundaries from the side of high refractive index and consequently do not undergo a phase change, so that the interference conditions are reversed and equation 3.2 now becomes the condition for constructive interference and a maximum of intensity for these rays. But, since ray 1 is much brighter than ray 2, ray 2 is much brighter than ray 3, and so on, the interference between rays 1 and 2 predominates and equation 3.2 remains the condition for a minimum of reflected intensity. If the film is of lower refractive index than its surrounding medium, ray 1 does not undergo the phase change but rays 2, 3, 4, . . ., do. Nevertheless, the same phase difference of π exists between rays 1 and 2 and, since these are the two brightest rays, equation 3.2 still applies as the condition for a minimum of intensity and equation 3.3 as the condition for a maximum. Since also rays 1 and 2 are dominant, the interference produced approximates to that of a two-beam system, for which the intensity distribution follows a cosine-squared law giving light and dark fringes of equal width.

The transmitted rays also interfere but there is now no phase difference introduced between the rays and therefore the interference conditions are reversed. Equation 3.2 now becomes the condition for a maximum and equation 3.3 for a minimum.

If the film is thick, the interfering rays appear well separated unless viewed along a direction nearly normal to the film. To observe interference a lens is used to bring the rays together at the eye, although this is not required if the film is thin or can be viewed along the normal. For nearly normal viewing $\theta = \theta' = 0$ and hence $\cos\theta' = 1$.

If monochromatic light from an extended source falls onto a thin transparent film, the resultant intensity produced by interference varies with the viewing angle θ. The extended source ensures that all values of the angle θ are available. The interference minima fall almost to zero for the reflected rays but not for the transmitted rays, since the minima are viewed against the bright source. If an extended white light source is used, the condition for maxima is satisfied at different angles by different wavelengths of light, producing different coloured maxima as the viewing angle is changed.

3.2 Interference by Wedge Films

The condition for a minimum due to interference between rays reflected from a thin film is, by equation 3.2,

$$2n't\cos\theta' = m\lambda$$

The thickness of this film is thus

$$t = \frac{m\lambda}{2n' \cos\theta'} \tag{3.4}$$

If now a wedge film of small angle α is considered, as in *Figure 3.2*, it is seen that, every time the thickness of the film changes by an amount Δt such that m in equation 3.4 changes by unity, the condition for a minimum occurs, giving a dark fringe. Thus a wedge film formed, for

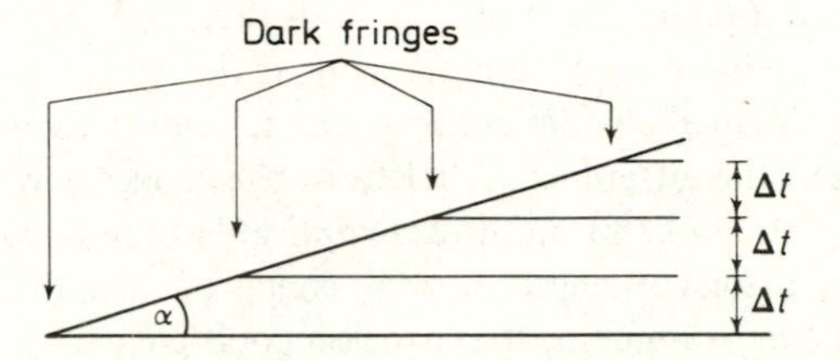

Figure 3.2. Formation of interference fringes by a wedge film

example, between two flat glass plates and illuminated by a monochromatic source appears as a series of dark fringes, lying parallel to the line of intersection of the plane surfaces and with a fringe being formed every time the wedge changes thickness by an amount

$$\Delta t = \frac{\lambda}{2n' \cos\theta'} \tag{3.5}$$

In practice the wedge is viewed normally so that $\cos\theta' = 1$ and generally the film is of air so that n' can be taken as unity. Equation 3.5 then becomes simply

$$\Delta t = \frac{\lambda}{2} \tag{3.6}$$

This enables thickness changes to be measured since each fringe is a line of constant film thickness and thus is analogous to a contour line on a map. It also enables angles to be accurately determined. For example, supposing the spacing of the fringes formed by a wedge film illuminated with mercury green light of wavelength $5{\cdot}46 \times 10^{-7}$ m is measured with a travelling microscope focused on the wedge film, then a dark fringe is formed each time the thickness of the wedge has changed by

$$\Delta t = \frac{5{\cdot}46 \times 10^{-7}}{2} = 2{\cdot}73 \times 10^{-7} \text{ m}$$

So, if there are 10 fringes in a distance of 25 mm, the thickness of the wedge has increased by 10 Δt and the angle of the wedge is

$$\alpha = \frac{10 \times 2{\cdot}73 \times 10^{-7}}{2{\cdot}5 \times 10^{-2}} = 0{\cdot}000109 \text{ rad}$$

$$= 22''$$

This method is used to test engineers' slip gauges. The gauge to be tested and the master gauge are 'wrung' down side by side into good contact with a flat plate. A flat glass plate is then laid across the gauges and illuminated with monochromatic light. If the gauges are of unequal height, a wedge shaped air film is formed and, from the spacing of the resulting fringes, the difference in height of the gauges may be calculated. An adaptation of this method is applicable to most problems involving comparisons of length or small changes in length as, for example, in the measurement of expansion coefficients.

3.3 Interference Topography

Interference provides the most important method for the testing and absolute measurement of flatness. A test plate is used as a standard; this is usually made from fused silica glass, which is hard and therefore resistant to scratching and wear. It also has a low coefficient of thermal expansion and is therefore less susceptible to local deformation owing to temperature gradients. These plates are usually made in sets of three. They are all worked together until each plate when matched against the other two produces a wedge film with straight fringes or, what is a more accurate comparison, produces a parallel film which in white light shows a single fringe of one colour over the whole field. This condition can be satisfied for two plates if they both have matching curved surfaces but it can only be satisfied for three plates if they are all flat.

The test plate is laid on the surface under test, which should be a reflecting surface free of any dust. The test plate is pressed down on one side to produce a wedge film and viewed with monochromatic light; any deviation from a straight line fringe shows that the test piece is not flat. It is easy to measure a deviation of 1/10 of a fringe spacing, which is equivalent to a height variation in the surface of $2{\cdot}7 \times 10^{-8}$ m with mercury green light. Whether the height variation is an elevation or a depression can be decided by observing whether the fringe deviation is towards or away from the direction of higher-order fringes.

The test plate need not be flat. Concave and convex plates are often used, for example, in the manufacture of ball-bearings and lenses. If the lens being tested and the test plate have the same radius of curvature, the parallel film of negligible thickness which forms when they are brought together shows a uniform interference colour in white light. A difference in radii produces a wedge film and circular fringes, the number of which may be used to deduce the difference in radii.

The method described for testing flatness can also be used to examine surface finish. Surface finish is also commonly measured by a stylus method which utilises the same principle as a gramophone pick-up. Variations in the topography of the surface under test cause vertical movements of the stylus, which may be electronically amplified and displayed using a chart recorder. Such methods have the disadvantages that the stylus itself may cause damage as it is drawn over the surface, that the examination is restricted to a line, and that a false or non-representative reading may be obtained, depending, for example, on whether the path traced by the stylus is parallel or normal to the finishing direction of the worked surface. Optical interference methods have none of these disadvantages: the test plate need not be in contact with the surface and an area is examined. However, to obtain comparable or greater sensitivity than the stylus method, only a small localised area may be examined at a time and consequently the two methods may be regarded as complementary.

For the general examination of a surface, a Fizeau interferometer as shown in *Figure 3.3 (a)* may be used. A small circular aperture at A at the focus of a well-corrected lens is illuminated by monochromatic light. Thus a parallel beam is produced which is normal to the reference test plate surface. Interference fringes, commonly termed Fizeau fringes, are then formed between this surface and the surface being examined. Light reflected back is deviated by a half-silvered plate and viewed by means of an eyepiece. If instead the aperture stop at A is removed and replaced by an extended monochromatic source, circular fringes will be formed when the two interfering surfaces are parallel. This is because all light falling on the surfaces at a particular angle will contribute to a particular fringe. That is, a cone of semi-angle θ' will contribute to a fringe of order m, according to equation 3.2. Other values of the angle θ' will satisfy this equation for other integral values of m. The circular interference fringes formed are termed Haidinger fringes and have ring diameters proportional to the square roots of the natural numbers. As the interfering beams are parallel, the fringes appear to be at infinity so that the eyepiece must be refocused from the Fizeau arrangement.

To obtain greater magnification, a microscope is required to examine the specimen surface and the fringes formed. One method is to use a special microscope objective, as in *Figure 3.3 (b)*, with a beam-splitting cube cemented to the front lens. Two prisms with their hypotenuse faces partially silvered are cemented together to form the cube. Rigidly

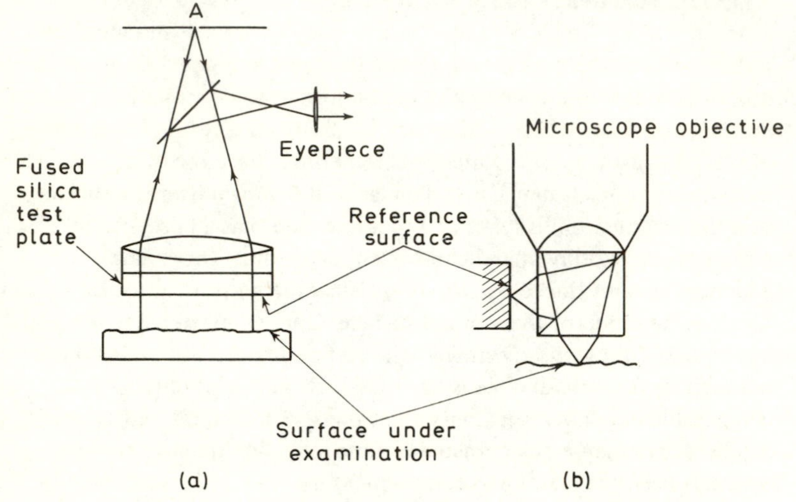

Figure 3.3. Methods for studying surface topography by interference

mounted to one side of the cube is a small reference surface which must be flat and free from any surface blemishes; it is therefore usually an optically worked glass surface with a thin layer of aluminium evaporated onto it. The position of the reference flat is such that it is always in focus. Thus, when the surface to be examined is brought into focus, its image is superimposed on the image of the reference surface. With monochromatic light, thin-film interference occurs. The specimen table can be tilted to produce a wedge film and straight line fringes or a parallel film and a uniform fringe. Any scratches or surface blemishes show up as distortion or displacement of fringes; these can be measured in terms of the fringe separation and hence the depth of the scratch calculated. The magnification of such objectives is governed by the working distance available to mount the beam-splitting device but, using various arrangements, objectives are available with magnifications up to ×40 – with a ×10 eyepiece this provides a total magnification of ×400.

This method is ideally suited for examining very fine finishes on small areas – for example, for the examination of razor blade edges and watch pivots – but it can also give much useful information by the localised examination of larger areas.

3.4 High-contrast Fringes

As was mentioned in Section 3.1, interference produced by reflections at a thin film in which the surfaces have a low reflectivity, as for example an air film between two glass plates, approximates to a double-beam interference system. With this type of interference the intensity distribution follows a cosine-squared law. Thus a deviation produced in a fringe owing to a varying film thickness caused perhaps by a small scratch can only be measured to an accuracy of about 1/10 of the fringe separation. If, however, the reflectivity of the surfaces is increased by coating them (using vacuum evaporation techniques) with, say, silver, multiple-beam rather than double-beam interference will be produced. Multiple-beam interference fringes may be made very sharp and fine, so that under *ideal* conditions a fringe displacement as small as 1/250 of the fringe separation may be resolved and measured. For example, using monochromatic mercury green light with a wavelength of $5{\cdot}46 \times 10^{-7}$ m, the fringe separation for an air film is equivalent to a thickness change of $2{\cdot}73 \times 10^{-7}$ m (equation 3.6). Thus a resolution of 1/250 of a fringe displacement makes possible the resolution of a thickness change in the film of 1×10^{-9} m. It must, however, be emphasised that this high resolution applies only to height variations from the mean surface and not to the linear dimensions of surface features. The resolution of the extent of surface features is governed by the resolution of the optical viewing system.

A reflected fringe system is generally found to be of most use as many specimens of interest are opaque. For example, a metal surface may be studied by laying over it a flat glass plate that has been partially silvered and observing the fringes produced by reflection. The dark fringes will then be narrow and sharp and the light fringes broad, that is, the fringe system will appear as dark lines on a bright background. This appearance will be reversed for fringes formed by transmission.

In *Figure 3.4* an incident ray of amplitude a undergoes a series of reflections and transmissions, where R is the fraction of the amplitude of the light reflected at a surface and T is the fraction transmitted. Then the sum of the reflected amplitudes is

$$\begin{aligned} A_R &= aR + aRT^2 + aR^3T^2 + aR^5T^2 + \ldots \\ &= aR + aRT^2(1 + R^2 + R^4 + \ldots) \\ &= aR + \frac{aRT^2}{1 - R^2} \end{aligned} \tag{3.7}$$

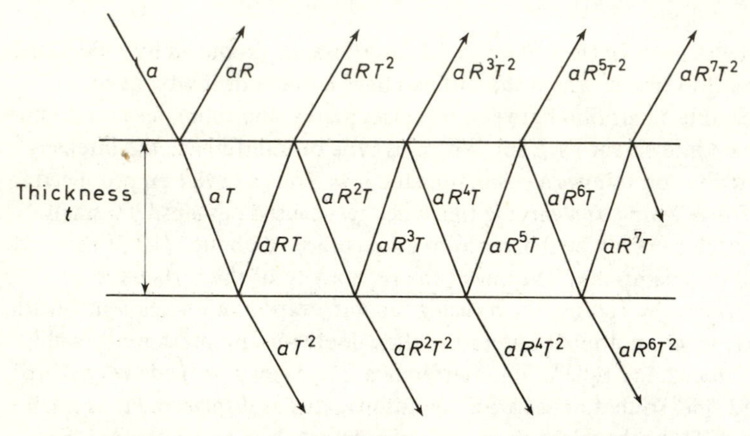

Figure 3.4. Multiple reflections at a thin film, showing the amplitudes of reflected and transmitted rays

by summing the geometric series. Similarly the sum of the transmitted amplitudes is

$$\begin{aligned} A_T &= aT^2 + aR^2T^2 + aR^4T^2 + \ldots \\ &= \frac{aT^2}{1 - R^2} \end{aligned} \tag{3.8}$$

With high-reflectivity films, that is, with R approaching unity, the sums of the amplitudes are the same apart from the extra ray of amplitude aR occurring in the reflected series. To calculate the intensity distribution of light within the fringes it is also necessary to take into account the phase difference between successive rays. It is found in practice that, with the surfaces coated with metal films for high reflectivity, the phase change introduced in a ray reflected at a glass-to-metal boundary is not very different from that introduced at an air-to-metal reflection. Assuming then that the same phase change is introduced in the first reflected ray as in all the succeeding rays, the resulting intensity of the transmitted light in a direction θ to the normal to the surface may be shown to be

$$I_T = \frac{1}{1 + [4R^2 \sin^2 (\delta/2)] / (1 - R^2)^2} \tag{3.9}$$

where

$$\frac{\delta}{2} = \frac{2 \pi t \cos \theta}{\lambda} \tag{3.10}$$

for a separation t between the reflecting surfaces. If there is no absorption within the film, then the reflected intensity is given by

$$I_R = 1 - I_T \tag{3.11}$$

For a highly reflecting surface, that is, with the reflecting power R^2 approaching unity, the term $(1 - R^2)^2$ in equation 3.9 will be very small. Consequently, on either side of an angle θ coinciding with a direction of maximum intensity, given by $2t \cos \theta = m\lambda$, that is, when $\delta/2 = m\pi$, the term $\sin^2(\delta/2)/(1 - R^2)^2$ will increase rapidly from zero. Thus the intensity I_T as given by equation 3.9 will fall off rapidly from a maximum value, so that the bright fringes formed by transmission through a film with highly reflecting surfaces will be much narrower than the dark fringes. That is, bright sharp fringes will be formed on a dark background.

As has already been shown, the sums of the transmitted and reflected amplitudes differ only by a constant term when highly reflecting surfaces are involved. Therefore the intensity distribution of the fringes produced by reflection will be the exact complement of that of the fringes produced by transmission, providing that there is no absorption within the film. The fringes produced by reflection will thus appear as dark sharp fringes on a bright background.

In practice, the metal layer deposited over the surfaces to increase the reflectivity also has an absorption effect. With increased absorption, the contrast between the bright and the dark fringes is reduced so that the fringe visibility is reduced. An optimum thickness of metal has therefore to be deposited that is sufficient to give high reflectivity and sharp fringes but not so thick that high absorption will destroy the contrast. The fringe width is determined only by the reflectivity of the deposited layers and is quite independent of the absorption.

It can also be shown that the fringe half width, that is, the width at half the peak intensity, as a fraction of the fringe spacing, is

$$\text{half width} = \frac{1 - R}{\pi\sqrt{R}} \times \text{(fringe spacing)} \tag{3.12}$$

Thus, with a reflecting power of $R^2 = 0{\cdot}8$, say, the half width of a fringe is approximately 1/28 of the fringe spacing. With mercury green light of wavelength $5{\cdot}46 \times 10^{-7}$ m, the fringe spacing is equivalent to a thickness change of $2{\cdot}73 \times 10^{-7}$ m so that the half width of the dark fringe is approximately equivalent to a thickness change of 1×10^{-8} m. Thus a deviation in a fringe equal to, say, 1/5 of the width of a fringe would be equivalent to a thickness change of 2×10^{-9} m. In comparison, using unsilvered glass plates, this same deviation of 1/5 of a fringe would correspond only to a thickness change of $2{\cdot}73 \times 10^{-7}/5 = 5 \times 10^{-8}$ m since the fringes formed are of equal width.

3.5 Lasers

Little has yet been said in this chapter concerning light sources, other than that a monochromatic source is required for optical interference systems. Of especial importance as a source of monochromatic light in many fields of research is the laser; it is not, however, particularly suitable for thin-film interferometry as for this an extended source is required. The laser, standing for 'Light Amplification by Stimulated Emission of Radiation', has, apart from its very high degree of monochromaticity, a feature that makes it very different from conventional sources such as a mercury vapour lamp. A conventional source emits incoherent radiation, that is, radiation that is spread over a range of frequencies and amplitudes with individual waves having different phases and polarisations. A laser, however, emits coherent light, that is, the radiation within a particular mode is all in phase, has the same frequency, and is emitted as a virtually parallel beam.

As electromagnetic energy passes through matter, interaction can only take place between the electromagnetic wave and the atoms of the medium if the frequency ν of the wave is equal to E/h, where E is the energy level to which an excited atomic system can be raised and h is Planck's constant, equal to $6{\cdot}626 \times 10^{-34}$ J s. The atoms of the medium will generally be in a mixture of ground and excited states, so that excited atoms will be emitting photons and returning to the ground state and ground state atoms will be absorbing photons of energy and becoming excited atoms. If now the medium consists mainly of excited atoms, then an incoming photon of the correct frequency will most likely strike an excited atom, causing that atom to emit a photon and return to its ground state energy. This emitted photon will have exactly the same phase and frequency as the incident photon. Both photons will then be free to strike other atoms and, providing there are only relatively few atoms in their ground states so that only a few photons will be absorbed, successive interactions will lead to a multiplication of the number of photons. Thus, to make a laser operate, the energy of the system must be raised so that there are far more atoms in an excited state than there are atoms in the ground state. The energy levels are then said to be 'inverted' and the injection of energy to produce this population invertion is termed 'optical pumping'.

In the ruby laser, chromium atoms are used as the active medium. About 0·05% of chromium is added to aluminium oxide to produce a

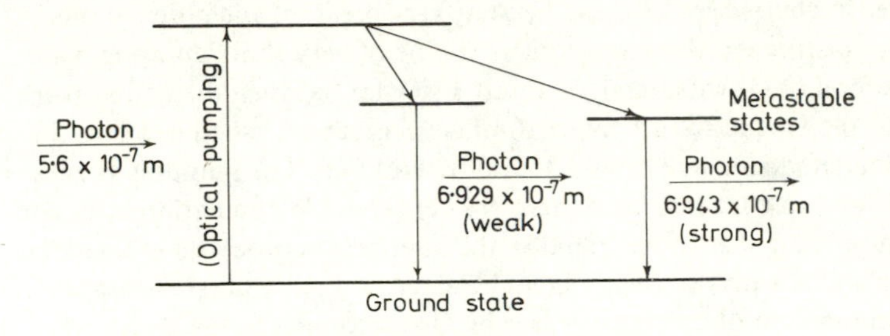

Figure 3.5. Energy levels involved in a ruby laser

doped ruby crystal of a light pink colour. An electronic flash tube, usually xenon, with a peak emitted wavelength of about $5{\cdot}6 \times 10^{-7}$ m, is used as a pumping source. An excited chromium atom then requires two steps to return to its ground state energy, as is shown in *Figure 3.5*. In the first energy drop, a small one, to one of two possible metastable levels, the energy released is absorbed by the crystal lattice, mainly in the form of heat. The atom can remain in this metastable state for perhaps several milliseconds after which it will drop to its ground state energy by emitting a photon. Depending on the particular transition, the emitted photon will have a wavelength of $6{\cdot}929 \times 10^{-7}$ m or $6{\cdot}943 \times 10^{-7}$ m. As the number of excited chromium atoms is raised by the repeated pumping action of the electronic flash, more and more chromium atoms will be emitting photons of wavelengths $6{\cdot}929 \times 10^{-7}$m and $6{\cdot}943 \times 10^{-7}$ m and the ruby crystal will begin to glow red. Once the number of atoms in the metastable states becomes significantly greater than the number in the ground state, the stimulated emission process begins to multiply rapidly, with nearly every emitted photon striking a chromium atom still in its metastable state, causing it to release another photon of the same wavelength. Owing to the greater population of the lower of the two metastable states, the light of wavelength $6{\cdot}943 \times 10^{-7}$ m is the most intense. If one end of the crystal (usually several centimetres long) is silvered to form a mirror and the other end partially silvered, photons moving parallel to the axis are reflected back and forth until the intensity of radiation becomes so great that the radiation can pass through the partially reflecting end face of the crystal as a pulse of virtually monochromatic coherent light lasting for a few milliseconds.

Using a technique known as *Q-switching*, the peak pulse power emitted by a ruby laser can become very great, of the order of megawatts, although this energy pulse will be of very short duration. One method of Q-switching is to insert a shutter between the end mirror and the crystal so that internal reflections cannot take place and the stimulated emission cannot therefore build up. The pumping process is allowed to continue until there is a considerable population inversion. On opening the shutter rapidly, the laser action proceeds as a sudden avalanche with the energy being released in a great burst. A simpler technique to obtain large pulses by Q-switching is to use a cell containing a solution of phthalocyanines between the end of the laser crystal and its mirror. This solution strongly absorbs the light emitted by the ruby laser so that the laser action is prevented from building up. However, as the pumping continues a time comes when there is sufficient intensity of radiation within the ruby crystal for some light to be reflected by the end mirrors and the laser action can just begin. This will slowly build up until at a certain threshold of intensity the phthalocyanine solution suddenly becomes bleached out, allowing full light transmission with its accompanying sudden avalanche of stimulated emission and the sudden release of the stored energy. The phthalocyanine solution then returns to its absorbing state and the process begins again. By using a continuous discharge mercury lamp rather than a flash tube, the ruby laser can be made continuously operating. The problem, however, is to obtain sufficiently good optical coupling to obtain a sufficiently strong pumping action. In practice this has been achieved by concentrating the light from the mercury lamp onto the laser with a system of concave mirrors and designing the laser with a conical end to condense the light radiation into a small volume within the crystal.

As well as solid-state lasers such as ruby and neodymium doped glass, which is tending to supersede the ruby laser, gas lasers are commonly used. Perhaps the commonest of the gas lasers is the helium–neon laser. Although working on the same principle as the ruby laser, it is very different in construction. The helium–neon gas mixture, in the proportions 10% helium, 90% neon and at a pressure of about 1×10^2 N m^{-2}, is contained in a quartz tube generally about $\frac{1}{4}$–$\frac{1}{2}$ m long and several millimetres internal diameter. The flat end windows are inclined at the 'Brewster' angle (Section 3.7) to polarise the emitted beam. The end mirrors, which are mounted externally to the gas tube, are usually made concave with a radius of curvature equal to their separation. With plane mirrors, difficulty is experienced in adjusting them to be accurately parallel so as to return the light back and forth along the laser tube axis but, with spherical mirrors, the problem is somewhat eased. The design of the mirrors also has an effect on the particular modes in which the laser operates. Diffraction effects are produced on reflection at the mirrors, resulting in some resonant reflections being just off axis. These slightly different resonance conditions can lead to the laser operating in a number of modes differing by a small fraction of the primary laser frequency. Within a mode, the frequency may be monochromatic to perhaps a few parts in 10^{14}, limited mainly by mechanical vibrations of the system. As before, one of the end mirrors is only partially silvered to allow the light to emerge. The mirrors in fact are not simply silvered but are of alternate layers of high and low refractive index material, each with an optical thickness of a quarter wavelength of the laser emission frequency and so designed to give high reflectivity, about 98–99%, at this specific wavelength only. Instead of a bright light source to produce excited atoms, a high-frequency electrical discharge is used as a pump.

Although basically similar in operation to the ruby laser, the helium–neon laser is slightly more complicated owing to the greater number of possible energy states. The helium atoms, excited by the electrical discharge, collide with the neon atoms and excite them, after which the helium atoms return to their ground state energy, to be re-excited by the electrical discharge. Owing to the impact within the discharge of electrons with the helium atoms, individual atoms are excited to one of two metastable levels. Collisions between these excited helium atoms and neon atoms result in neon atoms being raised to their 3s and 2s states, as in *Figure 3.6*.

An incident photon can excite some neon atoms from a 2p state to a 3s state and be absorbed, or can stimulate some neon atoms to pass from a 3s state to a 2p state with accompanying emission of another photon. The two photons will then be in phase and travel in the same direction. Coherent radiation of wavelength $6{\cdot}328 \times 10^{-7}$ m will then be emitted by the laser. Similarly, stimulated emission will be produced by the 2s to 2p transition. This radiation, owing to the multiplicity of possible transitions, will vary in wavelength over about $10–15 \times 10^{-7}$ m, with a peak emission at about $11{\cdot}5 \times 10^{-7}$ m. The transition from 2p to 1s gives rise to the incoherent red light characteristic of neon display tubes. Providing that

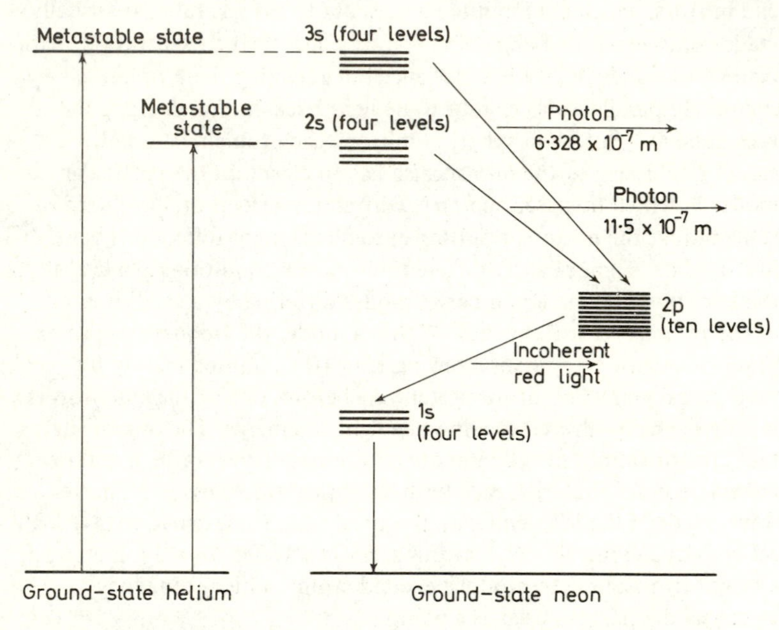

Figure 3.6. Energy levels involved in a helium–neon gas laser

most of the neon atoms are in the 2s and 3s states, most of the photons emitted during the energy drop to the 2p state will be used to trigger off the emission of photons from other atoms and the laser action can continue. The power emitted by a helium–neon gas laser is very small compared with that possible from, for example, a ruby laser and is generally of the order of milliwatts with perhaps 0·25 mW being typical for lasers in common use in laboratories, although output powers up to about 100 mW are possible.

The possible applications of lasers have been slow to materialise but lasers are now coming into use for a number of specialised purposes.

Most of the applications arise from it being a very monochromatic coherent source. This enables many types of interferometry systems to work over extended distances which are not possible using a conventional monochromatic source with its band of emitted frequencies. Thus the measurement of standards of length can be made with greater accuracy and ease, which in turn has importance in, for example, the calibration of surveying equipment. The high degree of parallelism of the beam is utilised in shadowgraph techniques by which the dimensions of components can be checked. For example, during the manufacture of steel rods, the diameter is required to be measured while the rods are still being made – while they are still hot and are vibrating. A laser beam can be scanned by means of a rotating mirror across a fixed volume that includes the rod to be measured. The light can be collected by a lens and projected onto a photocell, the output from which will be a function of the light not obscured by the rod and consequently will be a function of the diameter of the rod. Fire detectors can make use of the narrowness of the beam. For example, if a laser beam is directed at a distant photocell, any small deviation of the beam owing to varying refraction caused by hot air will move the beam off the photocell causing the fire alarm to sound. Surface-blemish detectors are used for searching for blemishes in paper during manufacture. By rapidly scanning the beam back and forth and measuring the reflectivity with a photocell as the paper passes, any blemishes are detected with a greater resolution than is achieved using conventional sources of light.

The power that can be obtained from lasers can be used for cutting and welding applications. Using commercially available CO_2 lasers, cutting speeds in excess of 3000 m min^{-1} can be achieved for paper and 1·3 mm thick steel sheet can be cut at 7·5 m min^{-1}. Drilling is usually performed with solid-state lasers since their pulsed operation makes them more suitable. Thus calcium tungstate–neodymium doped glass (Nd:$CaWO_4$) and yttrium aluminium garnet–neodymium doped glass (Nd:YAG) lasers are used for drilling holes in ruby bearings for watches. These holes are typically 0·01 mm in diameter and produced at the rate of six per minute. Although both pulsed and continuous lasers are used for welding, it is pulsed lasers that have proved to be most satisfactory. With these, either spot welds can be made or overlapping spots used to produce a continuous weld. The laser should be able to produce pulses between about 1–10 ms long and with peak powers of 10–100 kW. These conditions are necessary to prevent the surface vaporising and also to allow time for heat to be conducted to the lower levels of metal. Even so, welding is limited to a maximum depth of about 1 mm.

A number of specialised applications have also been found in the medical field, for example, in the treatment for a detached retina.

Finally, a warning must be given of the dangers of lasers. The primary danger is to the eyes as even very low-energy lasers can produce permanent damage to the retina. Care must therefore be taken to avoid any possibility of looking along the beam towards the laser. This may be done inadvertently during lining-up procedures or when there are any mirrors or reflecting metal surfaces in the path of the beam. There is also a danger with the high voltages involved with the high-frequency power supply.

POLARISATION

3.6 Polarised Light

Light is an electromagnetic radiation and as such is specifically a transverse wave. These waves arise from the vibrations of charged particles within the molecules that make up the light source, with vibrations of the correct frequencies producing electromagnetic waves in the form of visible light. The light wave sent out by one molecule may be considered to be plane polarised, since vibrations occur in only one plane containing the direction of propagation. For a large number of molecules, each of the vibrations is normally in a random direction and, as the resulting beam of light is made up of vibrations in all directions perpendicular to its direction of propagation, so that there is no preponderance of vibrations in any particular transverse direction, the light is said to be unpolarised. If, however, there is any asymmetry, the beam is said to be polarised.

Every electromagnetic wave has associated magnetic-field, electric-field, and electric-displacement vectors. The magnetic- and electric-field vectors and the direction of energy propagation, which in the case of light waves is the ray direction, are mutually perpendicular. In an isotropic material, such as unstrained glass, the electric-field and displacement vectors coincide. Thus light, being an electromagnetic wave, has associated with it these three vectors of which the electric-displacement vector, that is, the electric-vibration direction, is the most important.

Although two waves travelling along the same path but with their electric-vibration directions mutually perpendicular cannot produce interference effects, they can however interact to produce a radiation with special properties. The equations of the two electric-displacement waves may be written as

$$\xi_x = a \cos \phi \tag{3.13}$$

and

$$\xi_y = b \cos (\phi - \delta) \tag{3.14}$$

where ξ_x and ξ_y are the electric displacements in the *xz*- and *yz*-planes respectively at a time $t = \phi/\omega$, where ϕ is the phase and ω is the angular frequency. Both waves are travelling in the *z*-direction with the same velocity and frequency but with a constant phase difference δ and with different amplitudes, a and b, as in *Figure 3.7*. In the special case when $\delta = 0$ so that the waves are in phase, the resultant vibration may be represented by a vector of maximum length $(a^2 + b^2)^{1/2}$ and inclined at a constant angle $\tan^{-1} (b/a)$ with the x-axis, as in *Figure 3.7 (b)*. The length of the resultant vector varies with the phase ϕ. Thus the two mutually plane-polarised waves have combined to form a new plane-polarised wave.

If now the phase difference δ is taken as $\pi/2$, that is, the vibrations are in *quadrature*, equation 3.14 becomes

$$\xi_y = b \sin \phi \tag{3.15}$$

By plotting the displacements ξ_x and ξ_y at every instant as given by the phase ϕ and combining these displacements vectorially, it is seen that the resultant varies in both magnitude and direction. Combining equations 3.13 and 3.15, the resulting displacement is given by

$$\frac{\xi_x^2}{a^2} + \frac{\xi_y^2}{b^2} = 1 \tag{3.16}$$

and

$$\tan \phi' = \frac{\xi_y}{\xi_x} = \frac{b}{a} \tan \phi \tag{3.17}$$

Equation 3.16 is the equation of an ellipse, implying that the end of the resultant vector moves along an elliptical path, as shown in *Figure 3.7 (c)*. The light wave resulting from the combination of two

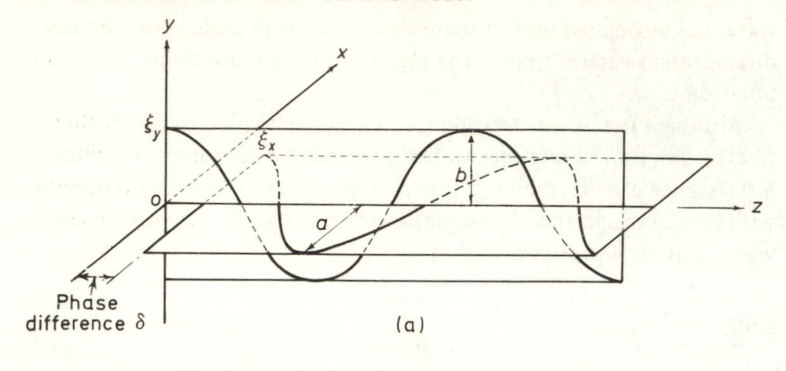

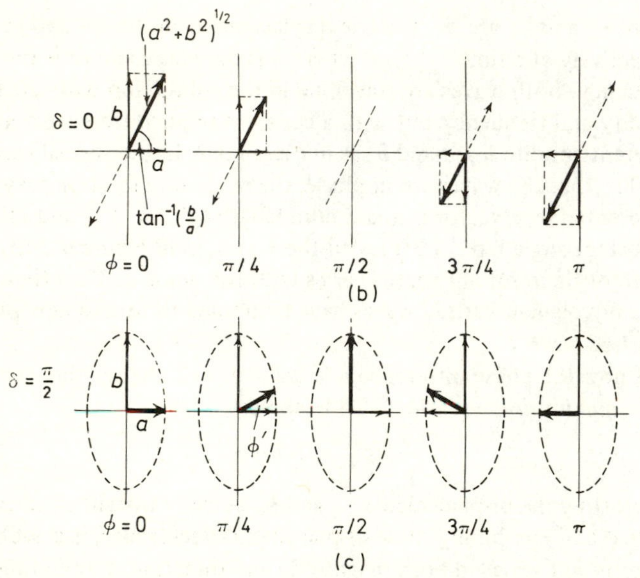

Figure 3.7. A representation of two mutually perpendicular waves of the same frequency (a) *and their vector addition to produce* (b) *plane-polarised and* (c) *elliptically polarised light*

mutually perpendicular plane-polarised waves in quadrature is thus described as *elliptically polarised* light. In the special case when the amplitudes of the combining waves are equal, that is, when $a = b$, equation 3.16 becomes the equation of a circle and the resultant light vibration is said to be *circularly polarised*. Increasing the phase difference δ to π again produces plane-polarised light but now with an inclination $\tan^{-1}(-b/a)$ to the x-axis. A further increase to $\delta = 3\pi/2$

reproduces the condition of equation 3.16, that is, elliptically polarised light again, except that now the inclination to the x-axis is ϕ' where

$$\tan \phi' = -\frac{b}{a} \tan \phi \tag{3.18}$$

Thus, looking towards the source of light, the end of the rotating vector rotates in a clockwise direction with respect to the observer as the phase ϕ progresses with time. The elliptically (or circularly) polarised light is then termed positive or right handed, whereas the direction of rotation defined by equation 3.17 is said to be negative or left handed.

For the more general case when the phase difference δ is not an integral multiple of $\pi/2$, the resultant vibration is still elliptically polarised but the major axes of the ellipse are no longer along the x- and y-axes. Circularly and plane-polarised light may thus be considered as special cases of elliptically polarised light.

3.7 Production of Plane-polarised Light

Any device that produces polarised light is termed a polariser; when this same device is used to detect polarised light it is then termed an analyser. A simple polariser for producing plane-polarised light involves reflection from, say, a glass plate. Unpolarised light incident upon a surface at an angle θ to the normal may be imagined resolved into two equal components, one of which has its electric-vibration direction parallel to the plane of incidence, that is, the plane containing the incident ray and the normal to the surface, whilst the other has its vibration direction perpendicular to the plane of incidence. Part of each of these components will be refracted at the surface and part reflected. For the fraction of each of the incident components reflected, that is, the reflection coefficients, Fresnel has shown that

$$R_{\mathrm{p}} = \frac{\tan(\theta - \theta')}{\tan(\theta + \theta')} \tag{3.19}$$

and

$$R_{\mathrm{s}} = \frac{\sin(\theta' - \theta)}{\sin(\theta' + \theta)} \tag{3.20}$$

where R_{p} and R_{s} are the reflection coefficients for the components with electric-vibration directions parallel and perpendicular to the plane of incidence respectively, and θ and θ' are the angles of incidence and refraction respectively. The angle of reflection is, of course, equal to the angle of incidence. These reflection coefficients are shown plotted in *Figure 3.8* for various angles of incidence θ for glass of refractive index 1·5. The negative reflection coefficient indicates a reversal of phase.

It will also be seen from *Figure 3.8* that the reflection coefficient R_p becomes zero for a particular value of the angle of incidence θ.

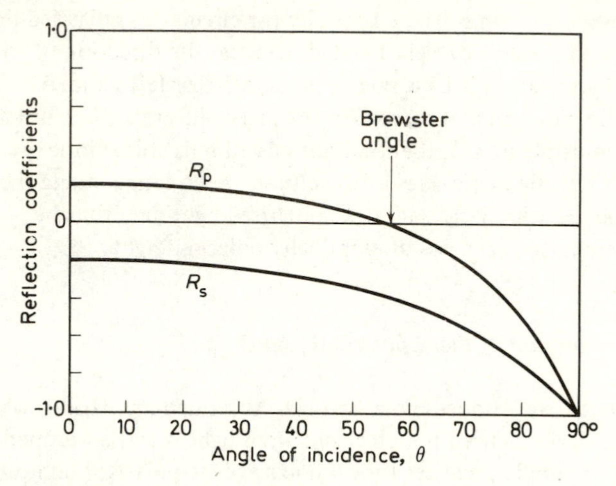

Figure 3.8. Variation of the Fresnel reflection coefficients with angle of incidence for an air/glass boundary (refractive index 1·5)

From equation 3.19 this will occur when $\theta + \theta' = \pi/2$, that is, when the angle of incidence $\theta = \tan^{-1} n$ where n, the refractive index, is equal to $\sin\theta/\sin\theta'$. At this angle, termed the Brewster angle, only that component with vibrations perpendicular to the plane of incidence is reflected. Thus, at this angle, for an air/glass boundary, 100% of the incident light component with electric vibrations parallel to the plane of incidence is transmitted, whereas about 15% of the perpendicular vibration component is reflected and 85% transmitted. This 15% that is reflected is pure plane-polarised light but, since it must be remembered that it constitutes only $7\frac{1}{2}$% of the *total* incident light energy, a single surface cannot be regarded as a very productive source of polarised light. However, by using a pile of glass plates inclined at the Brewster angle, as in *Figure 3.9*, the reflected plane-polarised beam may be sufficiently enhanced to provide a practical source. The transmitted beam will become progressively purer plane polarised as more of the perpendicular component is removed and reflected. Since the Brewster angle is a function of the refractive index, which in turn is a function of the wavelength, the plane-polarised reflected light condition can only be satisfied for one angle at one wavelength. A polariser consisting of a pile of plates, therefore, should only be used with monochromatic light.

Most methods of producing polarised light work on this same principle of producing unpolarised light, dividing it into two components having their electric vibrations at right angles, and then eliminating one of these components. Polarisation by dichroism involves this

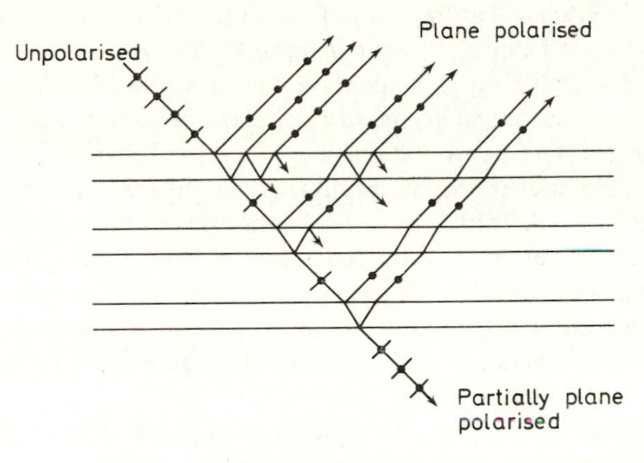

Figure 3.9. Production of plane-polarised light by reflection from a pile of glass plates

principle, by selectively absorbing one of the components. There are some naturally occurring crystals with this property, such as tourmaline and herapathite, but these tend to be small and not very good polarisers.

E. H. Land utilised this principle of selective absorption when he invented Polaroid. This is manufactured by stretching a transparent sheet of plastic, such as polyvinylalcohol, while warm. This aligns the long chain molecules parallel to the stretch direction. To prevent shrinkage the stretched sheet may be cemented to a rigid plastic sheet or, depending on the plastic used, it may be sufficient to allow it to cool after stretching. The stretched sheet is then dipped into iodine solution and iodine atoms diffuse into the plastic and arrange themselves along the long chains. After washing and drying, the sheet behaves as a polariser with the iodine chains acting as conducting 'wires'. The light component with electric vibrations parallel to the iodine chains will then induce an electric current in the conducting chains, which in turn produces heat. The energy of these vibrations will thus be absorbed. The other component, with electric vibrations perpendicular to the stretch direction, does not induce a current and so does not lose energy; it is therefore transmitted. The intensity transmitted must, of course, be less than 50% of the incident light and, with losses by reflection at the surfaces and absorption within the film, is usually less than 40%.

The third method of polarisation involves double refraction by an anisotropic medium. A ray of unpolarised or polarised light incident onto an isotropic medium such as glass obeys the laws of refraction, including Snell's law that the ratio of the sines of the angles of incidence and refraction is a constant for a particular wavelength and pair of media. If, however, a ray is incident onto an anisotropic medium, for example, a rhombohedral crystal of calcite, the refraction laws are not simply obeyed. The atoms that make up the crystal lattice of an anisotropic material are arranged differently in different directions through the crystal. Thus a ray of light inside the crystal 'sees' a different configuration of atoms depending on the direction in which it is travelling. Moreover, the electric vibrations that make up the ray of light interact differently with the atoms of the crystal lattice, depending on the relative orientation of the electric-vibration direction with the lattice.

A crystal of anisotropic material splits an unpolarised beam into two components that have their directions of electric vibration at right angles to each other. The different interactions of these vibrations with the crystal lattice generally cause the two components to take different ray paths. The two rays are both plane polarised. One ray, termed the *ordinary*, or *O-ray*, obeys the ordinary laws of refraction. This component always has its electric-vibration direction perpendicular to a particular direction in the crystal, called the *optic axis*. In crystals such as calcite there is only one direction of the optic axis and such crystals are therefore termed *uniaxial*. Another group of crystals having two optic axes, *biaxial* crystals, produce more-complicated optical effects. It should, however, be emphasised that the optic axis is a direction through the crystal, not a line as its name might imply.

The direction of electric vibration of the other component of the original ray, the *extraordinary* or *E-ray*, is inclined at an angle to the optic axis which depends on the crystal face and angle of incidence used. The E-ray does not obey the simple laws of refraction: it may not be refracted in the plane of incidence and does not obey Snell's law. It has different velocities through the crystal depending on its direction of travel, with its extreme velocity occurring when its electric-vibration direction is parallel to the direction of the optic axis. For sodium yellow light of wavelength $5{\cdot}89 \times 10^{-7}$ m, the refractive index for the O-ray in calcite is $n_O = 1{\cdot}658$ and the refractive index for the E-ray varies between 1·658 and an extreme value, $n_E = 1{\cdot}486$. The difference between the extreme values $(n_E - n_O) = 1{\cdot}486 - 1{\cdot}658 = -0{\cdot}172$. Calcite is therefore said to be a negative crystal, with a *birefringence* of 0·172 (for $\lambda = 5{\cdot}89 \times 10^{-7}$ m). Quartz, for example, is a positive crystal as $n_E > n_O$.

By isolating either the O-ray or the E-ray, a plane-polarised beam can be obtained with little wasted intensity and without any coloration. The rays can be isolated by means of a polarising prism. In this, one of the rays is removed by total internal reflection, leaving a single plane-polarised beam. The earliest design was a Nicol prism, which has now been superseded by such designs as the Glan Thompson prism. A piece of calcite is cut generally into a rectangular block, although the earlier designs of prism usually had a rhombohedral form, with the optic axis having a specified direction through the block, depending on the particular design. The crystal block is then cut in half along a diagonal, the cut surfaces polished flat, and the two halves cemented together with Canada balsam. Canada balsam has a refractive index approximately midway between the extreme refractive indices, n_O and n_E, for the calcite. The dimensions of the block are chosen such that one of the plane-polarised components will be transmitted through the block whilst the other is totally reflected at the calcite/balsam interface and thus effectively removed. The earlier design of Nicol prism had a side-to-length ratio of about $1{:}2\frac{1}{2}$ and an angle of useful aperture of about 22°. It also had a rhombic cross-section and slanting end faces which were a disadvantage, particularly the slanting end faces because they produced some elliptical polarisation. A Glan Thompson prism, however, has a rectangular cross-section with the end faces perpendicular to the lengthwise direction. It gives complete plane polarisation of the transmitted light and has a side-to-length ratio of about 1:3 with an aperture of about 30°. The Glan Thompson prism is about the best of the designs that have superseded the original Nicol prism.

3.8 Wave Plates

Wave plates or 'retarders' convert one form of polarised light to another. For example, a beam of plane-polarised light may be converted into circularly or elliptically polarised light, or vice versa, or the plane of polarisation may be rotated. Unfortunately, wave plates suffer from chromatism, that is, the magnitude of the change depends on wavelength. They should, therefore, only be used with monochromatic light.

Wave plates divide incident plane-polarised light into two components, slow down one component relative to the other – that is, they introduce a phase difference between the components – and then allow them to recombine. *Figure 3.10* shows a double-refracting plate cut with its optic axis as shown. An incident plane-polarised beam will be

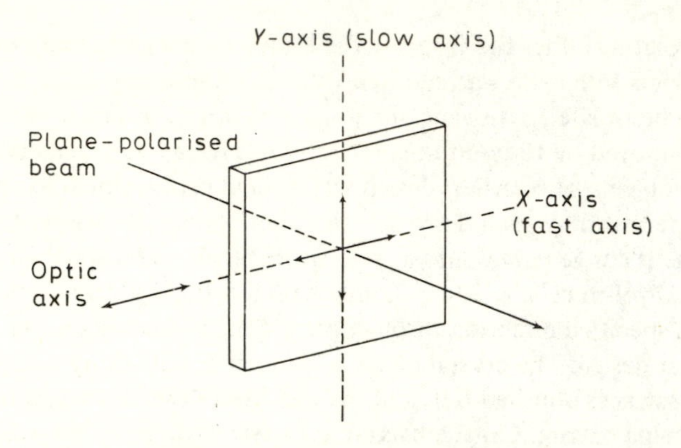

Figure 3.10. Calcite retardation plate

split into two components on entering the crystal plate, with the O-ray component having its electric-vibration direction perpendicular to the optic axis and the E-ray component having its electric-vibration direction parallel to the optic axis. The two rays then travel at different speeds. If the plate is made of calcite, the O-ray travels slower and the direction of its vibration is therefore termed the *slow axis*, as in the figure. The other axis is correspondingly termed the *fast axis*. In practice, calcite is so highly birefringent that the plates are too thin and fragile and therefore quartz or mica is used.

If the plate has a thickness t, the optical path difference introduced between the two components will be $t(n_O - n_E)$. The corresponding phase difference is thus

$$\delta = \frac{2\pi}{\lambda}\, t(n_O - n_E) \qquad (3.21)$$

This may be conveniently rewritten as

$$t = \frac{\lambda\delta}{2\pi(n_O - n_E)} \tag{3.22}$$

where t is the thickness of the plate required to produce a phase change δ using monochromatic light of wavelength λ. The two mutually perpendicular plane-polarised components may then be combined vectorially as in Section 3.6 and *Figure 3.7*. A plate that introduces a phase change of π, or a half-wave retardation, is termed a *half-wave plate*. A plane-polarised beam, incident normally onto a half-wave retardation plate with its plane of polarisation at an angle θ to one of

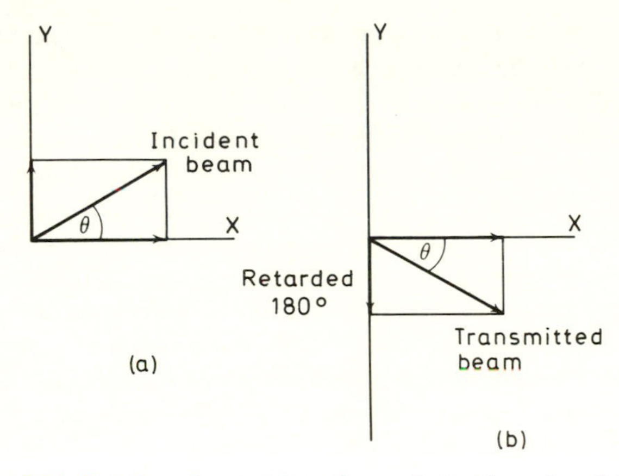

Figure 3.11. Rotation of an incident plane-polarised beam by a half-wave retardation plate

the plate axes, may be considered as made up of two components at right angles and of the same phase, as in *Figure 3.11 (a)*. After passing through the retardation plate, the slow ray has been retarded half a wavelength and is lagging π in phase; it may therefore be represented vectorially as in *Figure 3.11 (b)*, where the Y-component is shown rotated through 180°. Thus the resultant transmitted beam has had its vibration direction – that is, its plane of polarisation – rotated through an angle 2θ. In particular, a plane-polarised beam with its plane of polarisation inclined at 45° to one of the plate axes will be rotated through 90°. The plate will also convert right-handed circularly or elliptically polarised light into the left-handed circular or elliptical form, or vice versa.

Similarly a *quarter-wave plate* will introduce a phase difference of $\pi/2$. The transmitted beam is then the resultant of two mutually perpendicular vibrations which are $\pi/2$ out of phase. That is, the resultant vibration is an ellipse and becomes a circle only if the two components have equal amplitudes as when the plate axes make angles of 45° with the vibration direction of the incident plane-polarised beam. Conversely, a quarter-wave plate may be used to convert elliptically or circularly polarised light into plane-polarised light.

3.9 Applications of Polarised Light

Perhaps the most common application is that of polarising sunglasses. In sunlight the main source of light is overhead and the most brightly illuminated surfaces are horizontal. The bright surfaces are generally viewed obliquely, near the Brewster angle, so that the reflected light is partially plane polarised with its electric-vibration direction mainly horizontal. In polarising sunglasses the lenses are of a dichroic material orientated to absorb this horizontal vibration and to transmit only vertical components.

When the angle of reflection is nowhere near the Brewster angle, there is no polarisation by reflection. This situation is of especial interest, for example, to radar operators who are looking for faint spots on an oscilloscope screen. The screen also reflects light which tends to mask these faint spots. To stop these reflections, a circular polariser plate is used, consisting of a linear polariser and a quarter-wave plate together, in front of the oscilloscope screen. Light from the room passing through the plate is converted from unpolarised light into circularly polarised light, say, right-hand circularly polarised. This light is reflected from the oscilloscope screen but now as left-hand circularly polarised because the direction of circular polarisation is defined with respect to the ray direction. Therefore, on reversing the ray direction, the direction of circular polarisation appears reversed. This reflected left-hand circularly polarised light is then totally absorbed by the circular polariser plate because this transmits right-hand circularly polarised light. About 99% of the room light that was reflected from the oscilloscope screen is absorbed, whereas nearly half the wanted light from the screen is transmitted. In addition, to ensure that room light is not reflected from the circular polariser itself, it must be tilted at an angle to the viewing direction.

A method involving photoelastic analysis may be used in the design of complicated mechanical structures which are to be subjected to strain. In this method a model of the part to be studied is made from transparent plastic and placed between crossed polarisers. Polarisers are said to be crossed when the second one is rotated through 90° so as to extinguish the plane-polarised beam transmitted by the first. On looking through the assembly, only a dark field is seen. The plastic model is now subjected to forces similar to those which would be applied to the actual part under use but suitably scaled down. Light and dark coloured areas are seen in the model as molecules of the plastic are distorted by the applied stress and the plastic becomes birefringent, acting as a retardation plate. The amount of birefringence introduced at each point in the model is a measure of the strain occurring at that point and the two axes of the retarder at each point are parallel and perpendicular to the directions of the strain at that point. Thus the points of greatest strain are points of greatest retardation, resulting in the greatest mismatch of polarisation direction with that of the analyser and in most light being transmitted.

The measurement of the actual strain in a particular region may be made in several ways. Calibrated retarders may be used to cancel out the retardation introduced by strain in the model and hence give a measure of the retardation introduced. From subsidiary experiments on models of simpler shape the relation between strain and retardation, that is, the stress–optic constant, may be determined. Alternatively, it is possible to determine the retardation from the colour of the fringe at a particular point, since retardation depends on wavelength and the assembly may be illuminated by white light. This method, however, is at its most useful when, rather than for determining actual values, it is used to locate the points of maximum strain so that the structure may be strengthened at these points.

Photoelastic strain gauges are also used in the measurement of strain. The principle is the same as that for the photoelastic stress analysis method but, instead of making a model in plastic of the body under test, the plastic is coated onto the body as shown in *Figure 3.12*. The clear plastic is cemented into position with a reflecting cement. Since the light passes twice through the plastic, the sensitivity for a given thickness is doubled. As the body is strained, the plastic film also becomes strained and birefringent, giving a relative retardation between the two rays produced within the plastic. The amount of birefringence and consequent relative retardation is a measure of the strain, as just discussed. Photoelastic strain gauges have an advantage over the more usual resistance type strain gauges in that they examine the strain over an area rather than at a point. A disadvantage is that they can measure only surface strains that are also accessible to light.

Other materials show strain birefringence as, for example, glass. By examining glass between crossed polarisers it is easy to determine

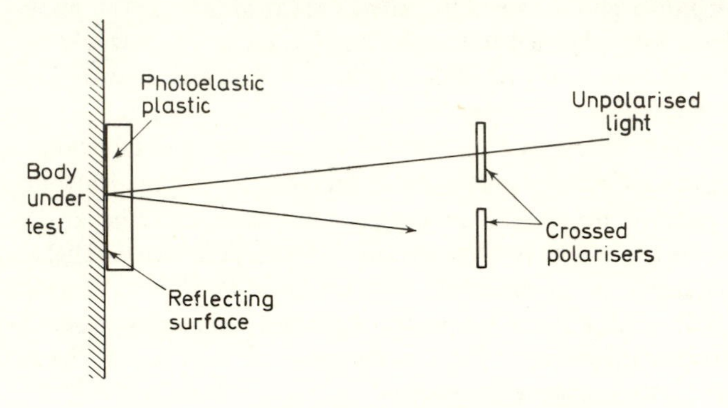

Figure 3.12. Use of a photoelastic strain gauge

whether it has been correctly annealed and hence is strain-free. If it is not, the strain may relieve itself in time by causing the glass to fracture.

A more specialised scientific use of polarised light is connected with *optical activity*. It is known that certain substances, for example, a sugar solution, have the property of rotating the plane of polarisation without changing the character of the linear vibrations. The rotation of the plane of polarisation depends on the distance travelled through the solution and the concentration of sugar. A polarimeter is an instrument for measuring this rotation. It contains a glass cell of known length and, with the solution in the cell, plane-polarised light is shone through it and the amount of rotation of the plane of polarisation is measured by a suitably sensitive analyser. The strength of the sugar solution may be determined if the instrument has previously been calibrated.

Two plane-polarisers inclined at a variable angle to each other may also be used to produce a variable density filter. *Figure 3.13* shows the transmission planes of a polariser and analyser, for example, two sheets of Polaroid, inclined at an angle θ to each other. Let A be the amplitude of the vibration transmitted by the polariser. This is resolved into two components at the analyser. One component, of amplitude A_1, has a vibration direction at right angles to the transmission plane of the analyser and is therefore absorbed. The other component, of amplitude A_2, has its vibration direction parallel to the analyser transmission plane and is therefore transmitted. This transmitted amplitude A_2 is equal to

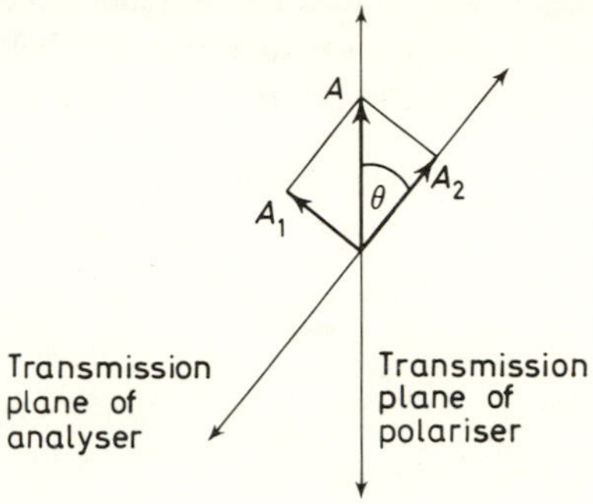

Figure 3.13. Amplitude of a plane-polarised beam transmitted by an analyser inclined to the polariser

$A \cos \theta$ and therefore the transmitted intensity, which is proportional to the square of the amplitude, varies as the square of the cosine of the angle between the transmission directions of the polariser and analyser.

3.10 Ellipsometry

Ellipsometry is a technique by which the thickness of thin films on metal may be determined, for example, a thin oxide film. The method is capable of determining thicknesses as small as 2×10^{-10} m but the computation is sufficiently laborious and complex to be impractical without the aid of a computer. The optical arrangement is shown in *Figure 3.14*.

To be able to determine the thickness it is first necessary to know the optical constants of the substrate. These are found using the arrangement shown in the figure together with a specimen free of any overlying film. The incident beam will in general be elliptically polarised and, by adjusting the relative orientations of the polariser and quarter-wave plate, the reflected light may be made plane polarised and extinguished by the analyser. The quarter-wave plate is first set with its fast axis at 45° to the plane of incidence. The orientations of the polariser and analyser, P and A respectively, are found for extinction.

The polariser is rotated through 90° and a second extinction position found, corresponding to an analyser orientation A'. All the angles are measured with respect to the plane of incidence with a clockwise rotation in the direction of propagation being regarded as positive. The reflected polarised light may be regarded as made up of two components, one with its electric-vibration direction parallel to the plane of incidence and the other with its vibration direction perpendicular. Then

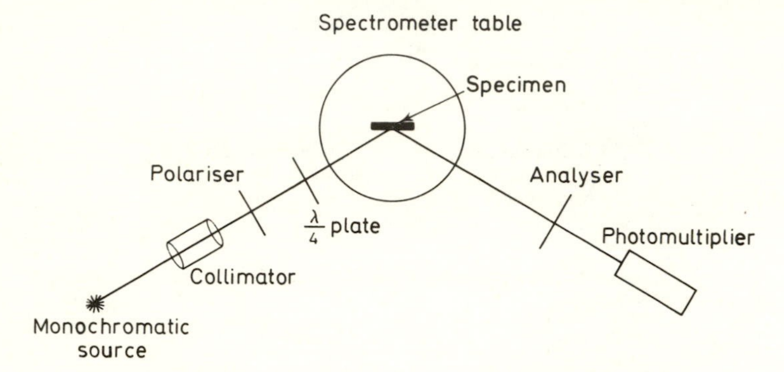

Figure 3.14. Optical arrangement for the determination of film thickness by ellipsometry

it may be shown that the relative phase change, Δ, between these two waves, on reflection is given by

$$\tan\Delta = \sin\delta' \cot 2P \tag{3.23}$$

where δ' is the retardation produced by the quarter-wave plate. Also, from

$$\tan^2\psi = -\tan A \tan A' \tag{3.24}$$

ψ can be found, where $\tan\psi$ is a measure of the change in the amplitude ratio of the two plane wave components on reflection. If ϕ is the angle of incidence of the beam on the specimen, the refractive index n and the absorption coefficient k of the substrate can be calculated from

$$n^2 - k^2 = \tan^2\phi \sin^2\phi \frac{(\cos^2 2\psi - \sin^2 2\psi \sin^2\Delta)}{(1 + \sin 2\psi \cos\Delta)^2} + \sin^2\phi \tag{3.25}$$

and

$$2nk = \tan^2\phi \sin^2\phi \frac{\sin 4\psi \sin\Delta}{(1 + \sin 2\psi \cos\Delta)^2} \tag{3.26}$$

The complex refractive index, $\hat{n}_3$, for the substrate is then given by

$$\hat{n}_3 = n - ik \tag{3.27}$$

The values of tan ψ and Δ for the substrate covered by the film under investigation are also determined for various unknown thicknesses of film.

It may also be shown that the ratio of the complex reflectances of the film for the two plane-polarised components is given by

$$\frac{\hat{R}_p}{\hat{R}_s} = \frac{\hat{r}_{12p} + \hat{r}_{23p}\exp(-2i\hat{\delta})}{1 + \hat{r}_{12p}\hat{r}_{23p}\exp(-2i\hat{\delta})} \cdot \frac{1 + \hat{r}_{12s}\hat{r}_{23s}\exp(-2i\hat{\delta})}{\hat{r}_{12s} + \hat{r}_{23s}\exp(-2i\hat{\delta})} \tag{3.28}$$

where the subscripts 1, 2, and 3 refer to the surrounding medium, film, and substrate respectively and the $\hat{r}$'s, the complex Fresnel coefficients, are defined by

$$\hat{r}_{12p} = \frac{\hat{n}_2 \cos\hat{\phi}_1 - \hat{n}_1 \cos\hat{\phi}_2}{\hat{n}_1 \cos\hat{\phi}_2 + \hat{n}_2 \cos\hat{\phi}_1} \tag{3.29}$$

$$\hat{r}_{23p} = \frac{\hat{n}_3 \cos\hat{\phi}_2 - \hat{n}_2 \cos\hat{\phi}_3}{\hat{n}_2 \cos\hat{\phi}_3 + \hat{n}_3 \cos\hat{\phi}_2} \tag{3.30}$$

$$\hat{r}_{12s} = \frac{\hat{n}_1 \cos\hat{\phi}_1 - \hat{n}_2 \cos\hat{\phi}_2}{\hat{n}_1 \cos\hat{\phi}_1 + \hat{n}_2 \cos\hat{\phi}_2} \tag{3.31}$$

$$\hat{r}_{23s} = \frac{\hat{n}_2 \cos\hat{\phi}_2 - \hat{n}_3 \cos\hat{\phi}_3}{\hat{n}_2 \cos\hat{\phi}_2 + \hat{n}_3 \cos\hat{\phi}_3} \tag{3.32}$$

where the p and s subscripts refer to the parallel and normal components of the reflected wave. The complex refractive index, $\hat{n}_3$, for the substrate has already been determined, $\hat{n}_1$, the complex refractive index of the surrounding medium, usually air, is known, and estimates can be made for the value of $\hat{n}_2$ for the film. Here $\hat{\phi}_1$, $\hat{\phi}_2$, and $\hat{\phi}_3$ are the angles of propagation of the reflected beam in the surrounding medium, film, and substrate respectively. These angles are also generally complex and are related by

$$\hat{n}_1 \sin\hat{\phi}_1 = \hat{n}_2 \sin\hat{\phi}_2 = \hat{n}_3 \sin\hat{\phi}_3 \tag{3.33}$$

$\hat{\delta}$ is a function of the film thickness and is given by

$$\hat{\delta} = \frac{2\pi t}{\lambda} \cdot \hat{n}_2 \cos\hat{\phi}_2 \tag{3.34}$$

where t is the film thickness.

Using a series of estimated values for the complex refractive index of the film, that is, $\hat{n}_2$, values of $\hat{R}_p/\hat{R}_s$ can be computed from equation 3.28 for a series of values of thickness t. $|\hat{R}_p/\hat{R}_s|$ then represents the relative amplitude change, $\tan\psi$, of the parallel and normal components of the reflected wave and $\arg \hat{R}_p/\hat{R}_s$ represents the relative phase change, Δ, of these components. The series of computed values can thus be plotted in the complex plane in the form $\tan\psi$ against Δ for each estimated value of $\hat{n}_2$, with the thicknesses t being marked off along each curve.

The experimental values of $\tan\psi$ and Δ determined for a series of thicknesses of the film are also plotted and compared with the series of theoretical curves. A match between the experimental curve and one of the theoretical curves indicates that the complex refractive index of the film is that of the theoretical curve. The thickness of an individual film is then determined directly from the appropriate refractive index curve.

Four

MICROSCOPY

OPTICAL MICROSCOPY

4.1 The Optical Microscope

In a microscope, the objective which is a compound lens system is used to produce a magnified, real, inverted image. This image is then further magnified to produce a virtual image at the near point – that is, at about 250 mm from the eye, as in *Figure 4.1*. Since both the objective and eyepiece are compound lens systems, the rays are drawn with reference to the principal planes. Thus, rays may be drawn from a small object of height y to pass through the first and second principal foci F_O and F_O' of the objective to form a real image of height y'. The image of height y' is then further magnified by the eyepiece, which behaves as a simple magnifier. If the final image is to be formed at infinity, y' will be located at the first principal focus F_E of the eyepiece. In practice, the microscope is normally adjusted to form the final image at the near point, taken to be about 250 mm from the eye. This means that the image y' is located just inside the principal focus F_E of the eyepiece.

The lateral magnification produced by the objective is thus

$$M_O = \frac{y'}{y} \simeq \frac{l}{f_O} \tag{4.1}$$

where f_O is the focal length of the objective and l is the optical tube length. The *optical tube length* is defined as the distance between the back focal plane of the objective and the first focal plane of the eyepiece, which is approximately equal to the distance to the intermediate image y'. The eyepiece behaves as a simple magnifier and hence has a magnifying power equal to

$$M_E = \frac{D}{f_E} \tag{4.2}$$

where D is the least distance of distinct vision and f_E is the focal length of the eyepiece. The total magnification of the microscope is then $M_O \times M_E$. Microscope objectives are normally manufactured with powers of x5, x10, x20, x40, and x100 and eyepieces with powers of x5, x10, and x15. The magnification quoted by the manufacturer should

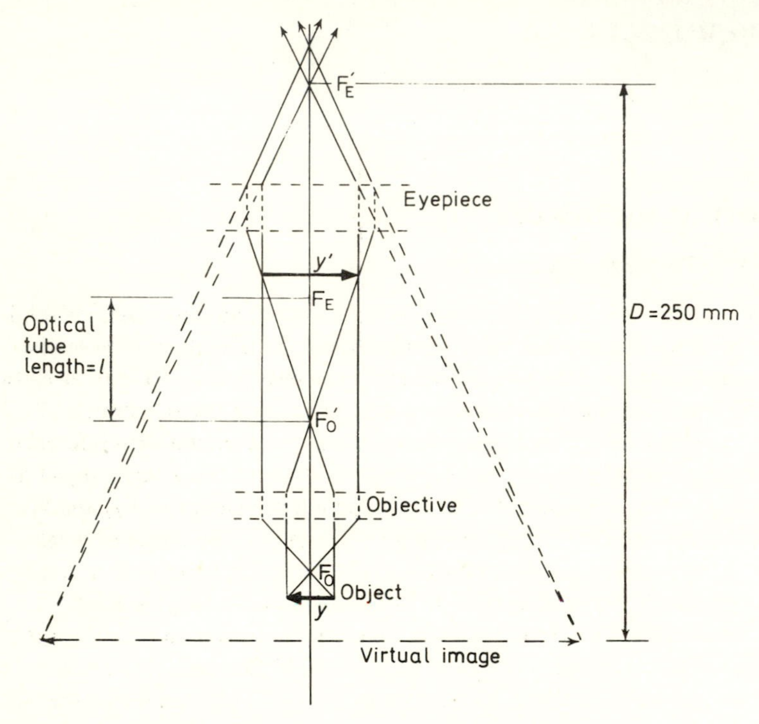

Figure 4.1. Constructional ray diagram for a microscope

be regarded as a nominal value. Therefore, if the magnification is required to be known with greater accuracy so that the size of an object may be measured, a divided graticule is located at the plane of y' and an object of known length is used. This length can then be measured in terms of the graticule divisions and hence the graticule can be calibrated.

The *mechanical tube length* is the actual length of the microscope tube as measured from the top, where the shoulder of the eyepiece rests, to the bottom, where the shoulder on the screw-in objective locates. This has been standardised at 160 mm. Generally the mechanical and optical tube lengths are approximately the same. The thread on the objective mount has also been standardised by the Royal Microscopical Society. Since the diameter of the eyepiece is also standardised, most eyepieces and objectives by different manufacturers should be interchangeable.

The *parfocal distance* is the distance from the object to the bottom of the microscope tube where the objective fits. This distance must be constant so that on changing an objective to one of a different power the object will be kept approximately in focus. This distance, however, is not standardised, so that objectives supplied with different microscopes will not be parfocal if mixed indiscriminately. To get the back

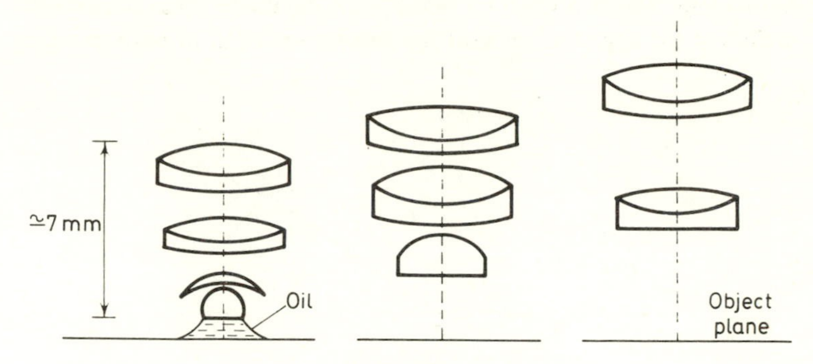

Figure 4.2. Arrangement of lenses in ×100, ×40, *and* ×10 *microscope objectives*

focal plane of the objective in the correct position with reference to the microscope and the position of the object, the lenses are suitably positioned within the objective jacket.

Owing to the highest possible resolution being required, a microscope objective must be as free as possible from aberrations. The design problems are, however, aggravated by the very steep angles involved and the large curvature of the lens surfaces necessary with high magnifaction, up to ×100, and the requirement of a high light-gathering power. *Figure 4.2* shows typical lens arrangements for high-, medium-, and low-power objectives. With a ×100 objective, the working distance, that is, the distance between the first lens surface and the object, is about 0·25 mm. Consequently much of the light coming from the object will meet the first lens surface at grazing incidence and will thus be lost by reflection instead of being refracted into the objective. To ensure that a wide angle cone of rays may enter the objective to give maximum light intensity and resolution, it is necessary to have a film of oil of the same refractive index as the objective's first lens so as to bridge the gap between the object and the first lens surface, that is, the object space and the first lens are made optically homogeneous. The oil used is generally cedar oil of refractive index about 1·517 for sodium yellow light. After use an oil-immersion objective must be cleaned by wiping off the oil with lens tissue; occasionally extra cleaning may be required and in this case the tissue may be moistened with xylene, but never alcohol.

This oil film is also important in reducing the aberrations of the objective in the following way. It can be shown that the condition for the formation of a perfect image of a small lateral object centred on the axis of a lens system is given by Abbe's sine condition: that the ratio of the sines of the angles made with the axis by each pair of corresponding

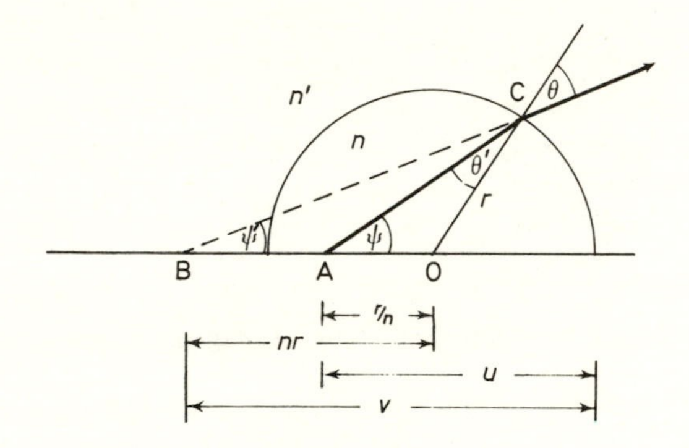

Figure 4.3. Aplanatic points of a sphere

object and image rays is constant, irrespective of the sizes of the angles involved. If a lens obeys this condition, then the image formed will be free from any spherical aberration. If the condition is also satisfied for an off-axis object so that the lateral magnification for all zones of the lens is constant, then the lens will also be free from coma. If an image free from both coma and spherical aberration for all zones can be formed for one particular object point, the system is said to be *aplanatic*. In practice, however, this term is often applied if these aberrations are eliminated for just one zone. In particular, a sphere is truly aplanatic for a particular pair of conjugate object and image points; these points are shown in *Figure 4.3* where the sphere has a radius r and refractive index n, surrounded by a medium of refractive index n'. By application of the thin-lens formula

$$\frac{n'}{v} - \frac{n}{u} = \frac{n' - n}{r} \tag{4.3}$$

with the New Cartesian sign convention, and where v and u, the image and object distances, are as shown in the figure, it is seen that an object point at A such that AO = r/n gives rise to an image at B, such that BO = nr. Also, from the figure,

$$\frac{\text{BO}}{\text{OC}} = n = \frac{\text{OC}}{\text{AO}} \tag{4.4}$$

and since ∠BOC = ∠AOC the triangles BOC and COA are similar. Therefore $\psi' = \theta'$ and $\psi = \theta$. Then

$$\frac{\sin \psi}{\sin \psi'} = \frac{\sin \theta}{\sin \theta'} = \frac{n}{n'} \tag{4.5}$$

by Snell's law. Equation 4.5 also shows that the ratio of the sines of the angles ψ and ψ', the angles made with the axis by a pair of corresponding object and image rays, is a constant. That is, the Abbe sine condition is satisfied and the sphere is aplanatic for the two points A

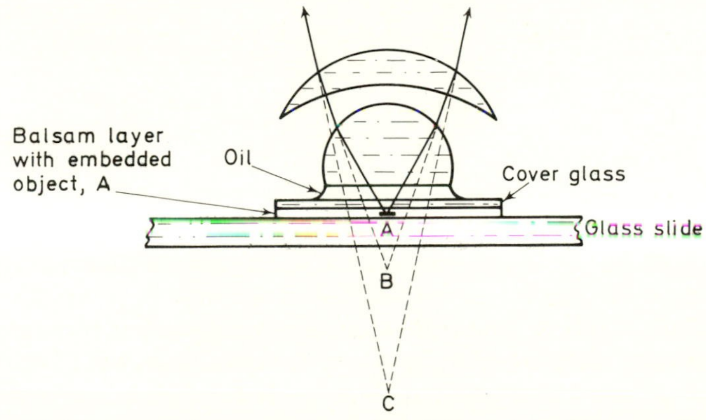

Figure 4.4. An Amici front component of an oil-immersion microscope objective

and B. This property of the sphere is utilised in the front lenses of high-power microscope objectives. The object cannot, of course, be embedded in a glass sphere but, by grinding part of the sphere away and using a drop of oil of the same refractive index as the glass between the glass and the specimen, the required optical condition can be achieved.

If the rays diverging from the image at B are in turn incident on a meniscus lens as in *Figure 4.4*, in which the centre of curvature of the first surface is at B, these rays will not be deviated since they are incident normally to the surface. If also the second surface of the meniscus lens is of such a radius of curvature that B is at its first aplanatic point, these rays will be refracted again without spherical aberration or coma and appear to come from the second aplanatic point, C, for this surface. In such an arrangement, which follows the design first suggested by Amici, magnification is thus achieved by both lenses aplanatically. Unfortunately this process cannot be carried on further owing to the increasing amount of chromatic aberration. Later lenses in the system are then used to give greater magnification and to compensate for the chromatic aberration introduced by the Amici front

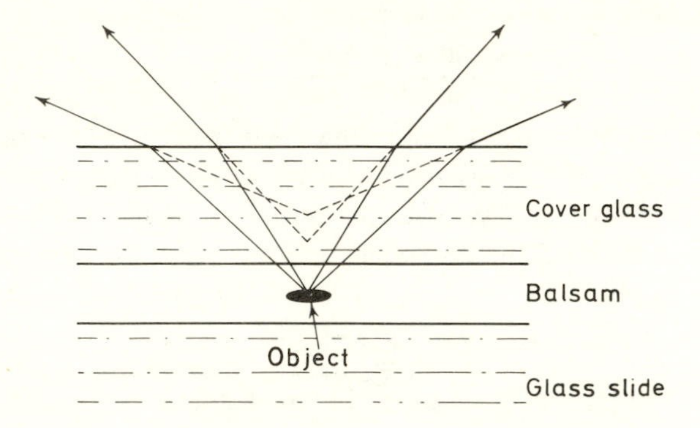

Figure 4.5. Spherical aberration introduced by a cover glass

and the oil, which has a different dispersion to the glass. Medium-power lenses, such as the ×40 shown in *Figure 4.2*, do not use oil immersion and since, for practical reasons, the first surface must be plane, the front lens cannot be completely aplanatic. The middle and back lens components again are used to give further magnification but also have to compensate for the spherical aberration and chromatic aberration introduced by the front.

Biological and other delicate specimens are usually embedded in Canada balsam and protected by a cover glass as in *Figure 4.5*. Canada balsam is used since it has a refractive index approximately the same as that of the glass. Unfortunately, however, the parallel cover glass and balsam layer introduce spherical aberration. It is necessary, therefore, at least for the higher powers, to have different objectives corrected for use with covered or uncovered objects. To correct for variations in cover glass and balsam thickness, the draw tube, which is the tube into which the eyepiece drops, is made movable so that the mechanical tube length may be changed from its nominal 160 mm; by shortening the tube length, for example, the spherical aberration resulting from too thick a cover glass can be compensated. With a ×100 objective corrected for a standard cover-glass thickness of 0·18 mm, a variation in cover-glass thickness of 0·03 or 0·04 mm will give a noticeable deterioration in the image unless compensated for by adjusting the tube length.

Having produced a magnified image of an object with an objective lens, an eyepiece is then required to magnify this image further. Two types of eyepiece are in general use. These are the Huygenian and the Ramsden types and they have already been described in Section 2.9. There is, however, a limit to the magnification that should be used. Beyond a certain magnification, set by the resolution in the instrument, no further information is added to the image. It is therefore important to the microscopist to know the highest resolution that can be obtained.

4.2 *Resolving Power of a Microscope*

Owing to the method of illumination of an object under a microscope, the object can to a first approximation be regarded as being self-luminous. Thus, if two neighbouring small objects are viewed through a microscope, their images are in fact two circular diffraction patterns produced by the circular apertures involved. In *Figure 4.6*, *a* is the

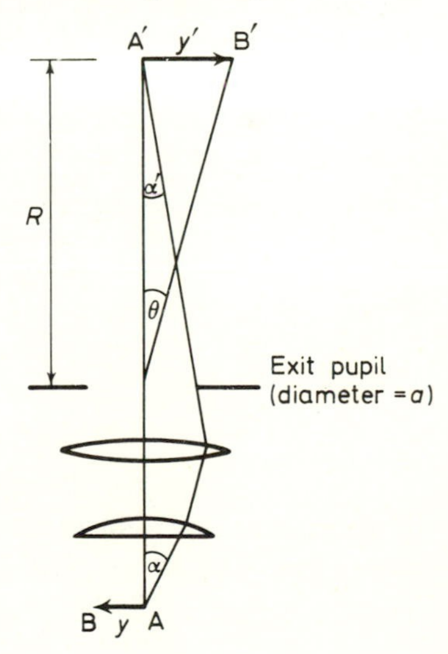

Figure 4.6. Two small self-luminous point objects at A and B are imaged by a microscope as two circular diffraction patterns centred at A′ and B′

diameter of the exit pupil of the objective and is a distance R from the objective image plane. Since the distance R is large compared with the distance between the objective exit pupil and the objective or the object plane, this exit pupil may be regarded as the aperture which produces the diffraction. Two small self-luminous point objects A and B situated a distance y apart are then imaged as two circular diffraction patterns centred at A′ and B′, a distance y' apart. According to the Rayleigh criterion for resolution, the point objects A and B are only just resolvable as two point objects if the central maximum of one diffraction pattern, say at A′, coincides with the position of the first diffraction minimum of the other pattern centred at B′, as in *Figure 4.7.*

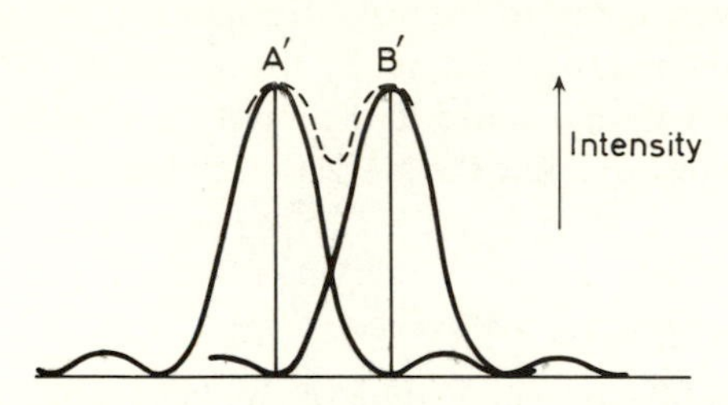

Figure 4.7. Rayleigh criterion for resolution

That is, the distance $A'B' = y'$ must be equal to the radius of the central bright disc, termed Airy's disc, of either of the circular diffraction patterns, for A and B to be just resolvable. The angular radius of the Airy disc is given by

$$\sin\theta = \theta = \frac{1{\cdot}22\,\lambda}{a} \tag{4.6}$$

where θ is small. Also, from Figure 4.6,

$$\theta = \frac{y'}{R} \tag{4.7}$$

and therefore

$$y' = \frac{1{\cdot}22\,\lambda R}{a} \tag{4.8}$$

If the extreme rays through the objective make angles α and α' with the axis (*Figure 4.6*), then $\alpha' = (a/2)/R$ and

$$y'\alpha' = 0{\cdot}61\;\lambda \tag{4.9}$$

The Abbe sine condition for perfect image formation for a particular wavelength λ may be written as

$$ny \sin \alpha = n'y' \sin \alpha' \qquad (4.10)$$

where n is the refractive index of the object space (about 1·5 for an oil-immersion objective) and $n' = 1$ where the image space is in air. Thus, the distance between the objects A and B for them to be just resolvable is,

$$y = \frac{y' \sin \alpha'}{n \sin \alpha}$$

$$= \frac{y' \alpha'}{n \sin \alpha} \qquad (4.11)$$

since α' is small. That is

$$y = \frac{0{\cdot}61\,\lambda}{n \sin \alpha}$$

$$= \frac{0{\cdot}61\,\lambda}{\text{NA}} \qquad (4.12)$$

where NA is the *numerical aperture*. In deriving this expression for the resolving power, it was assumed that the point objects A and B were self-luminous. This, however, is not strictly true. Abbe examined the problem and taking into account that the objects were illuminated by a condenser concluded that the factor of 0·61 in equation 4.12 should more correctly be 0·5 and therefore resolving power

$$y = \frac{0{\cdot}5\,\lambda}{\text{NA}} \qquad (4.13)$$

Assuming that the smallest angle that can be resolved by the eye under the normal illumination conditions of microscopy is 2′ (0·00058 rad), then two point objects which can only just be resolved with the microscope when they have a separation y will, on magnification, have an apparent separation $y' = 0{\cdot}25 \times 0{\cdot}00058$ m (*Figure 4.8*). The total magnification produced by the microscope is then

$$M = y'/y = (0{\cdot}25 \times 0{\cdot}00058) / \frac{0{\cdot}5 \times 5 \times 10^{-7}}{\text{NA}}$$

$$= 580\ \text{NA}$$

$$= \text{say, } 600\ \text{NA}$$

to allow for some spare magnification and where the mean wavelength of light is taken as 5×10^{-7} m. The numerical aperture NA is equal to $n \sin \alpha$. Sin α cannot be greater than unity and with oil immersion – that is, $n = 1{\cdot}5$ – the largest NA normally available is about 1·3, using a ×100 objective. This gives the total magnification required to just resolve the two point objects as approximately 1000 times, which is obtainable with a ×100 objective and a ×10 eyepiece. If the power of

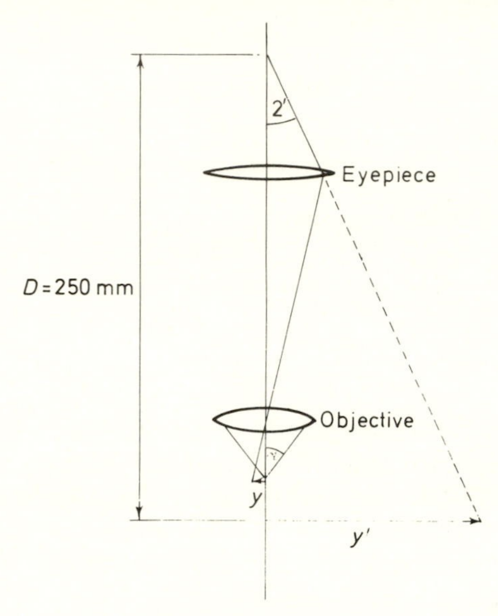

Figure 4.8. Illustrating the limiting resolution of a microscope

the eyepiece is increased, no more information is obtained and the image is less bright. The extra magnification is then termed 'empty' magnification.

Abbe has also shown that the resolution depends on the NA of the condenser used to illuminate the object. To realise the full resolution capabilities of the objective it is, therefore, necessary to use a condenser system that is sufficiently well corrected and with an NA comparable to that of the objective.

4.3 Condenser Systems for Microscopy

To illuminate the object and provide bright field illumination conditions, a condensing system is required to concentrate the light.

This system should be capable of focusing light from an external source to a point focus so that a cone of rays may diverge to fill almost the whole aperture of the objective uniformly with light. *Figure 4.9 (a)* shows the arrangement of an Abbe condenser which has a maximum NA of about 1·2 when used with oil immersion. A drop of oil is placed on top of the condenser which is then raised to meet the underside of the glass slide on which the specimen is mounted. Such an arrangement is, of course, only required for use with a x100 oil-immersion objective so that the space between the top lens of the condenser and the bottom lens of the objective is completely filled with oil, glass, and the specimen

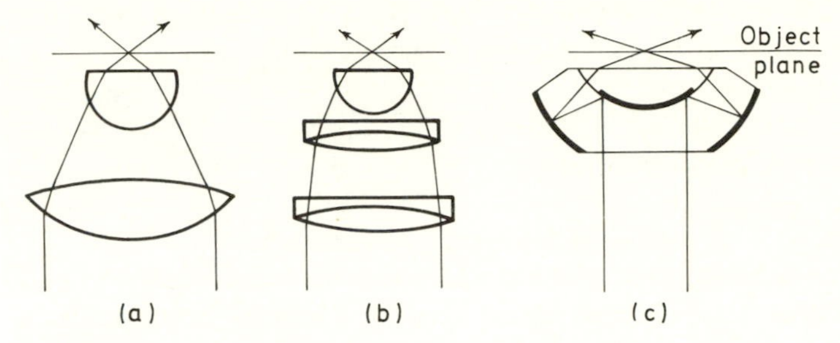

Figure 4.9. (a) Abbe, (b) achromatic, and (c) dark-field condensers

under examination – there is no air layer. For lower-power objectives with lower numerical apertures, the system is used dry, that is, without oil. Some Abbe condensers are made with three lenses, with the top lens being removable for use with low-power systems. With an Abbe condenser, however, owing to its comparative optical simplicity, there is considerable spherical and chromatic aberration, although in the three-lens Abbe arrangement the spherical aberration may be reduced considerably. These aberrations may be corrected in an achromatic condenser [*Figure 4.9 (b)*] in which the design requirements are similar to those of the objective but the manufacturing tolerances are not so stringent.

An iris diaphragm fitted below the condenser allows the numerical aperture to be adjusted to suit the objective being used. If the aperture of the condenser is reduced so that the NA of the illuminating cone is less than the NA of the objective, then the resolving power of the objective will be reduced. However, some reduction of the condenser aperture usually leads to an improvement in image contrast. Thus, depending on the nature of the object under investigation, a compromise has to be made between contrast and resolution.

If a condenser of greater NA than that of the objective is used, then, by stopping out a central cone of rays equal in NA to that of the objective, direct rays can be prevented from entering the microscope. However, if the field to be magnified is made up of a number of small objects capable of scattering, diffracting, or refracting the light in any way, then some of this light may enter the objective. These small objects are then seen as bright objects on a dark background instead of the more usual converse. The objects are made far more visible but they may be difficult to identify owing to diffraction effects. Besides a simple stop on the axis of the condenser, various systems have been designed to give oblique illumination without the wastage of light caused by a blank stop, as, for example, the design shown in *Figure 4.9 (c)*.

4.4 *Setting-up Procedure*

If the full capabilities of the microscope are to be realised, then it is most important that all the optical components are in alignment and the illumination is correctly adjusted. To adjust the components for alignment it is usual to provide some type of centring arrangement for the condenser and its attached iris diaphragm, so that the condenser may be brought into alignment with the axis of the eyepiece and objective. An approximate method of alignment is to set up the microscope for low-power use, using a x10 objective, with the substage mirror adjusted to reflect light into the microscope tube as in *Figure 4.10*. It should be noted that some microscope substage mirrors are plane on one side and concave on the other. Whenever a condenser is used the plane mirror must be utilised; the concave side is only used for low-magnification work when a condenser is not necessary.

The microscope tube is now raised so that the objective is focused on a position about 30 mm above its normal position. An image of the aperture iris will then be seen when looking through the microscope. This image should be centred in the field of view by adjusting the substage centring screws. The mirror is then tilted to give a central illumination of this image.

A more accurate method of alignment requires the lamp and mirror to be adjusted first. With the condenser, objective, and eyepiece removed from the system and with a piece of ground glass or tissue paper resting on top of the microscope tube, the position of the lamp and mirror should be adjusted to direct the light up the tube to give a centrally positioned patch of light on the ground glass screen. The condenser, objective, and eyepiece are now returned to their correct positions and the microscope focused on a specimen. The vertical position of the condenser is then adjusted to focus an image of the lamp iris diaphragm, the field iris, in the plane of the specimen. This image will then appear superimposed on the image of the specimen. If the image of the field iris does not appear central in the field of view,

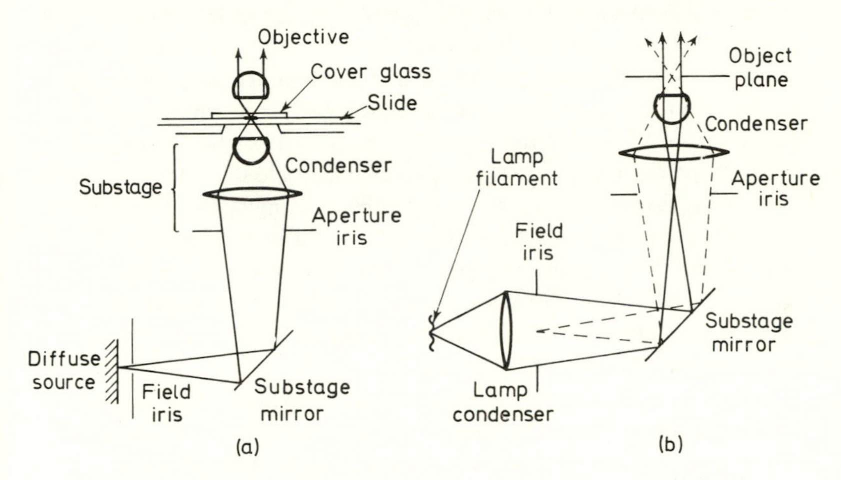

Figure 4.10. Illumination system for a microscope to give (a) critical and (b) Köhler illumination

this is corrected by centring the substage unit holding the condenser and aperture iris. The position of the mirror should not be altered.

With the optical components now in alignment, the required method of illumination may be chosen and the microscope adjusted for optimum viewing conditions. Providing the substage centring screws are not interfered with, the alignment of the condenser, objective, and eyepiece will, of course, remain in adjustment. What must now be adjusted is the focus of the lamp, if it has a focusing system, and the opening of the field and aperture irises. Two types of illuminating conditions are in common use. These are *critical illumination* and *Köhler illumination.* For critical illumination the source of light is usually an opal lamp, or could be any diffuse source such as daylight, whereas Köhler illumination requires a source that can be focused to a small intense spot of light (*see Figure 4.10*). Critical illumination is suitable for low magnification but Köhler illumination is to be preferred in general because it gives a greater intensity of light at the object and is therefore the most suitable for high magnification. It also gives a more uniform illumination and is therefore to be preferred for photomicrography.

With critical illumination the microscope, which is already in optical alignment, is focused on the specimen. The condenser is moved along the optic axis to focus onto the same plane as the specimen an image of the source and with it the field iris, so that the images of the specimen, source, and field iris will be superimposed. The tilt of the mirror is then adjusted to bring the image of the light source and field iris centrally into the field of view and the aperture of the field iris adjusted so that the light source only just fills the field of view. For low-power use it may be found preferable to defocus the condenser slightly to destroy the image of the source structure. The eyepiece should then be removed and looking down the microscope tube the opening of the aperture iris adjusted so that the back lens of the objective is just completely filled with light. This will give maximum resolution viewing but generally it will be found better to close the aperture iris until only about two-thirds of the aperture of the objective is filled with light, to improve the contrast at the expense of resolution.

With Köhler illumination the microscope is focused on an object as before and the condenser focused to produce an image of the field iris superimposed on the object. This image should be centred as before by adjusting the tilt of the mirror. The lamp condenser must now be focused to form an image of the lamp filament in the plane of the aperture iris, which will also be the position of the first focal plane of the substage condenser, as in *Figure 4.10 (b)*. The field of view should then be evenly illuminated. If it is not, then centring of the bulb filament by means of the lampholder centring screws is required. The field iris should then be closed down until it just encloses the field of view as before, and the opening of the aperture iris adjusted to fill between two-thirds and the whole of the aperture of the objective as viewed down the microscope tube with the eyepiece removed. This last adjustment is to give the required compromise between contrast and resolution. For a different magnification, requiring a different objective to be used, the openings of the field and aperture irises must be readjusted.

Although only transmission microscopes, that is, ones in which the illumination passes through the object, have been described, metallurgical microscopes are in common use. These are for use with opaque specimens and make use of a vertical illuminator. Light is brought into the system through a side tube between the eyepiece and objective and reflected along the microscope axis by a half-silvered plate. Light thus passes down through the objective, which acts as a condenser, to illuminate the specimen and is then reflected back through the objective and passes straight through the half-silvered plate to the eyepiece. An alternative arrangement for use with opaque specimens is to invert the optical system completely, as in *Figure 4.11*. This has the advantage

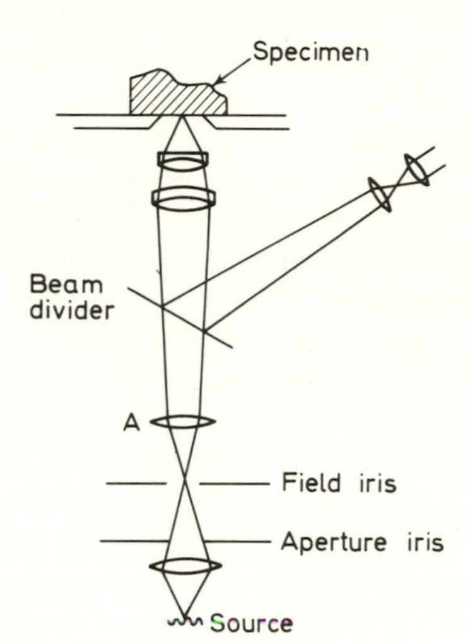

Figure 4.11. An inverted microscope arrangement for examination of opaque specimens

that only the side of the specimen being examined need be flat and the specimen can be of any thickness. The lens A in the figure serves to make the plane of the field iris conjugate with the image plane of the objective and the plane of the aperture iris conjugate with the exit pupil of the objective, allowing both field and aperture control to be exercised independently by the respective irises. Owing to the number of back reflections in the system, it is necessary for the lenses of objectives used with vertical illumination to have anti-reflection coatings. This is especially necessary since the general light level is low owing to absorption by the half-silvered beam divider and the generally low reflectivity of the specimen.

4.5 Polarised-light Microscopes

Essentially a polarising microscope is a conventional bright-field microscope with the addition of a polariser and an analyser. Earlier microscopes used calcite polarising prisms, such as the Nicol or Glan Thompson discussed in Section 3.7, but modern microscopes use Polaroid type filters. The polariser is located below the substage condenser and allows only plane-polarised light to enter the system, whilst the analyser is located above the objective, generally just below the eyepiece, and is used to detect the polarised light. If the analyser and polariser are crossed, that is, the analyser is orientated to extinguish the plane-polarised light produced by the polariser, then any depolarisation or rotation of the plane of polarisation produced by the object under examination will destroy the extinction condition in a localised region. Similarly any birefringence within the object, as is commonly the case in mineralogical specimens, will cause that region of the object to act as a retardation plate, as discussed in Section 3.8, giving coloration due to interference effects. Thus an almost featureless object, perhaps a slice of rock ground down in thickness to become virtually transparent or perhaps a fibrous specimen, will show up between crossed polarisers as an object with light and dark features that may also be coloured.

To achieve the full benefit of this type of viewing system, a number of microscope design features become of importance. The most important of these is the necessity for strain-free condensers and objectives. Any strain in these components will partially destroy the extinction condition of the crossed polarisers, giving light patches in an otherwise dark field. Therefore optical components for use with a polarising microscope should be specially selected. It is also necessary to make one of the polarising filters, usually the substage polariser, rotatable to give bright-field conditions when required, for example, when generally examining the specimen to select areas for more detailed study. To orientate the specimen with respect to the plane of polarisation produced by the substage polariser, the microscope should also be equipped with a rotating stage. This must be accurately centred so that the axis of rotation of the stage coincides with the optical axis of the microscope, otherwise an object in the centre of the field will not remain in the centre as it is rotated. The rotating stage is usually fitted with a degree scale enabling rotations to be measured. If small crystalline specimens are to be examined, it is generally essential to have a universal stage to enable the specimen to be tilted as well as rotated.

Some specimens show up differently in parallel or convergent illumination and it is therefore desirable to be able to switch easily between these two types of illumination. This is usually done by having a condenser with swing-out lenses. Other desirable features of a polarising microscope are that it should have a keyed eyepiece – so that it cannot be rotated – which is fitted with a cross-line graticule showing the orientations of the transmission planes of the polariser and analyser; and below the analyser should be a slot to allow the insertion of calibrated retardation plates into the beam. These can be quarter- or half-wave plates (Section 3.8) or wedges, and are used to compensate for and thus measure the retardation produced by the specimen. For more accurate measurement of retardation, a Berek compensator should be used: this is a small calcite plate which may be tilted by a measured amount. The plate is cut so that its optic axis is normal to the plate and thus introduces no retardation when in a horizontal position. As it is tilted a retardation is introduced which is a function of the angle of tilt, which may be calibrated to read retardation directly.

Further information concerning the specimen may be obtained by viewing its effect on light of varying obliquity. This is done by switching in a small lens, called a Bertrand lens, below the eyepiece, which enables the polarised-light pattern at the exit pupil of the objective to be viewed, rather than the specimen itself. This is termed conoscopic observation in contrast to the usual orthoscopic observation. Conoscopic observation enables, for example, the angle between the optic axes of a biaxial crystal to be measured. A universal stage is now even more necessary, so that the specimen can be suitably orientated and accurate angular measurements made.

4.6 Phase-contrast and Interference Microscopy

Most living cells or tissue show virtually no contrast when viewed under normal bright-field microscope conditions and therefore it is necessary to stain the specimen in some way to enhance the contrast between its different parts. This, however, may kill the cell or tissue or inhibit its natural functions. There is therefore a need to be able to examine the specimen without resorting to staining. Both phase-contrast and interference microscopy make this possible. Both systems are dependent on slight variations of refractive index within the specimen, which differs from the refractive index of the surrounding medium.

In the phase-contrast arrangement, a clear annular ring located at the first focal plane of the condenser is imaged in the rear focal plane of the objective, as in *Figure 4.12*. Thus, without an object in position, all the hollow cone of rays that pass through the clear substage annulus will pass through an annular region of the phase plate. If now an object is introduced, then some of the rays will be diffracted away from the ray paths shown in *Figure 4.12* and will consequently pass through a different region of the phase plate and not through the annular region. Depending on the refractive index of each part of the object, there will be a varying optical thickness which will introduce corresponding phase

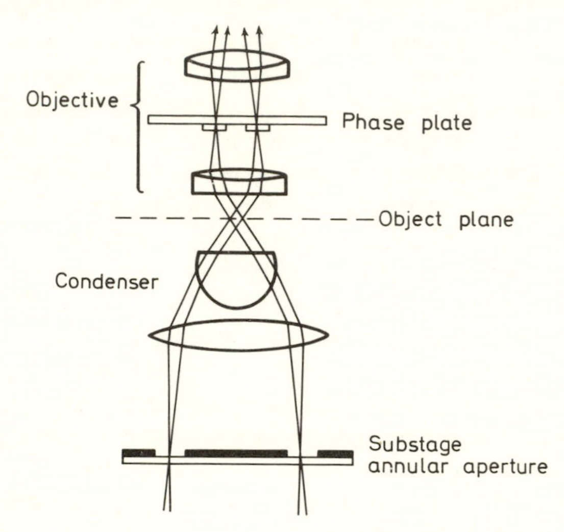

Figure 4.12. Arrangement of annular aperture and phase plate in a phase-contrast microscope

differences in the light rays. That is, the object is regarded as a phase object rather than the usual amplitude object.

A small object will produce a diffraction pattern consisting of a central maximum of intensity, the zero order, surrounded by a series of subsidiary maxima, of which the first-order maximum has the greatest intensity. The zero order, arising from the undiffracted rays, will fall on the annular region of the phase plate whilst the rays making up the first- and higher-order subsidiary maxima will fall outside this region. On average, the rays forming the first-order diffraction maximum are $\pi/2$ in phase, that is, a quarter wavelength, behind those making up the central maximum. If now the phase plate is coated with a layer that increases to π this retardation of the first-order diffracted rays relative to those of the zero order, then the zero- and first-order diffracted rays will be out of step and will destructively interfere. Thus the diffracting object will appear dark against a bright field when surrounded by a lower refractive index medium. In practice, the annular region has deposited on it, by vacuum evaporation, a coating of magnesium fluoride and a metal. The plate is then protected by cementing on a glass cover. The thickness of the magnesium fluoride annulus is such that its optical thickness is a quarter wavelength less than that of the surrounding cement region, causing the zero-order rays to be a half wavelength in advance of the diffracted first-order rays. The metallic layer is to introduce some absorption so that the amplitudes of the zero-order rays are reduced to be comparable to the amplitudes of the first-order rays and allow the destructive interference to be more nearly complete. This arrangement is termed 'dark' or 'positive' contrast because the phase object appears dark. 'Bright' or 'negative' contrast is the reverse of this and is produced by having a phase annulus that retards the zero-order rays by a quarter wavelength and brings them into step with the first-order rays, resulting in constructive interference and the object appearing bright against a dark background.

Interference microscopy works on very similar principles but does not use the object to introduce the separate interfering beams. Light passing through the object undergoes a retardation relative to light passing through a neighbouring clear region. On combining these separate beams, interference takes place and path differences are converted into intensity differences. There are various systems for producing the two separate beams. Perhaps the simplest in concept uses two matched objectives – one focused on a flat reference mirror and the other on the object under examination. The separate images are then superimposed and interference takes place. An approximately similar arrangement has already been described in Section 3.3 for examining opaque objects. Here only one objective was used owing to sufficient working depth being available with the low-power objective to allow the beam splitter to be incorporated. This arrangement is not practical for higher powers; using two objectives overcomes this difficulty but the higher the power the more prone the system is to vibration problems.

A more stable design is to use double-refracting components to produce the two interfering beams. One design, for example, uses two calcite plates. One of these is located above the condenser to convert incident plane-polarised light into two mutually perpendicular plane-polarised beams, one of which can pass through the object and the other through a neighbouring region, and the other calcite plate is below the objective and serves to bring the two beams back into coincident paths, as in *Figure 4.13*. The image formed is thus due to the superimposition of two different images of the object area, which are then interferometrically compared. The different retardations introduced by the specimen will depend on refractive index which is

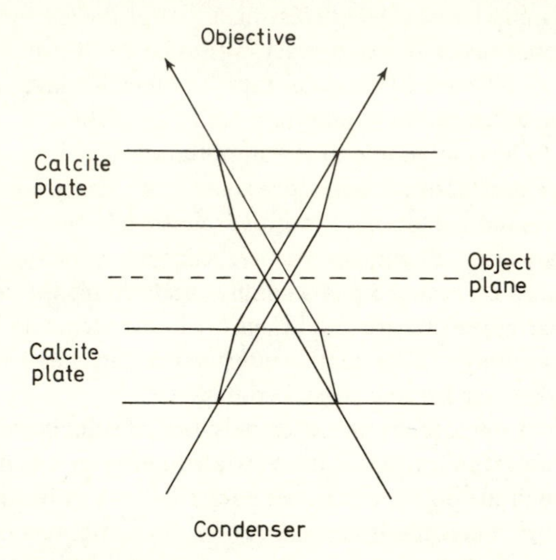

Figure 4.13. Arrangement of calcite plates to produce sheared wavefronts in an interference microscope

also a function of wavelength. Thus the image will be coloured which in turn will enhance the contrast.

Before the two mutually perpendicular plane-polarised beams can be made to interfere, however, they must be converted to circularly polarised beams by a suitably orientated quarter-wave plate. The resultant of these beams of opposite rotation, a plane-polarised beam whose orientation depends on the phase relationship between the two circularly polarised beams, is then viewed through an analyser located below the eyepiece.

4.7 *Photomicrography*

To obtain a photograph of the magnified image, a camera must be fitted to the microscope. The usual type of camera attachment is made to clamp onto the top of the microscope tube. Some designs of camera include a beam splitter and a supplementary eyepiece that allows the object being photographed to be viewed at the same time. With this design, an image that appears in focus through the eyepiece should also be in focus at the film plane of the camera. However, the eye is capable of accommodating some degree of out of focus but this is not acceptable for the image formed on the film and therefore considerable care is needed in focusing. The eyepiece is usually fitted with some form of graticule, such as a cross-line, and the microscope focused so that there is no parallax between the image as seen in the eyepiece and this graticule.

The image field in many microscope systems is slightly curved and therefore if it is sharply in focus at the centre of the field it will be slightly out of focus at the edges of the field. To accommodate this curvature on a flat photographic film or plate, a point a little way off-centre may be taken as the point for sharp focusing so as to obtain a more acceptable photograph. To check that the focal plane of the eyepiece, as defined by the eyepiece graticule, is optically coincident with that of the camera, a piece of ground glass should be placed in the film or plate position in the camera so that the ground side is towards the incoming rays. The microscope should then be focused to form a sharp image on this ground glass screen. To reduce the accommodating effect of the eye, a magnifying glass can be used to view the ground glass screen and the sharpness of the focus. If the image seen through the supplementary eyepiece is not then also sharply in focus, adjustments should be made to the position of this eyepiece.

The other problem of photomicrography is determining the exposure time which from a practical point of view should be as short as possible to reduce the effects of vibration. The illumination system therefore should be carefully adjusted to obtain the optimum conditions as, in particular, any unevenness in the illumination is especially noticeable in this type of photography. With a little experience, based on a few test photographs, reliable estimates of the exposure times can soon be made. If the conditions are changed, however, it should be remembered that the exposure time varies as the square of the magnification and inversely as the square of the usable numerical aperture of the illuminating cone of light. It also varies inversely as the ASA rating of the film. Further control is given by varying the brightness of the source, since an increase in the voltage of a filament lamp by 25% will approximately halve the required exposure time. However, varying the brightness of the source in this way will affect its colour temperature which may become of importance when using colour film. Generally a blue or 'daylight' filter is required to improve the colour balance. Exposure meters are also available for use with microscope cameras.

ELECTRON MICROSCOPY

4.8 The Electron Microscope

The importance of an electron microscope is not so much the very high magnifications that are attainable but the very high resolutions coupled with these magnifications, so making them 'useful' magnifications. It was shown in Section 4.2 that the resolving power of an optical microscope is limited to approximately half the wavelength of the light used and it is therefore necessary to decrease this wavelength before any improvement may be gained. Little, however, can be gained by mere reduction to, say, ultraviolet light wavelengths as in fluorescent microscopy, owing to the increased manufacturing difficulties partially offsetting the possible gain in resolution. To make any significant gain in resolution it is therefore necessary to use a radiation of significantly shorter wavelength, as for example electron radiation. This is more practicable than using, say, X-rays, as electrons can be focused by electric and magnetic fields whereas X-rays can only be focused by techniques involving reflection at glancing incidence.

Electrons have both particle- and wave-like properties. The ability of an electron microscope to form an image of an object depends on the particle-like properties of the electron, since it is these properties that determine the trajectories through the electric and magnetic fields. The wave-like properties, however, are of importance in relation to the resolution of the microscope. As was first postulated by de Broglie, the wavelength associated with an electron is

$$\lambda = \frac{h}{mv} \tag{4.14}$$

where h is Planck's constant, $6{\cdot}626 \times 10^{-34}$ J s, m is the mass of the electron, and v its velocity. This velocity is acquired by accelerating the electron through a potential difference E, which gives it a kinetic energy

$$\tfrac{1}{2}mv^2 = Ee \tag{4.15}$$

where e is the charge on the electron, that is, $1{\cdot}602 \times 10^{-19}$ C. From equations 4.14 and 4.15 the wavelength can be determined as a function of the accelerating voltage E. Owing to the high velocities involved, the relativistic change in mass should be taken into account, that is, the electron mass m is given by

$$m = \frac{m_0}{[1-(v^2/c^2)]^{1/2}} \tag{4.16}$$

where m_0 is the rest mass of the electron, $9{\cdot}108 \times 10^{-31}$ kg, and c the velocity of light in vacuum, $2{\cdot}998 \times 10^8$ m s^{-1}. In practice, however, the obtainable resolution is limited by the difficulties in design and manufacture of the electrostatic and magnetic focusing lenses, which limit the resolution of commercial instruments to a minimum resolvable distance of about $2–3 \times 10^{-10}$ m. With these instruments, magnifications up to about 500000 times are generally possible, depending on the particular design. However, instruments with resolutions of perhaps $10–20 \times 10^{-10}$ m and lower magnifications are adequate for many purposes and are additionally much lower priced than the highest-resolution instruments. Although much larger electron microscopes can be built with much higher accelerating voltages and higher resolutions, the cost tends to become prohibitive making them a non-commercial proposition. A high-magnification instrument must also be capable of producing a range of magnifications to deal with various problems and should also be capable of very low magnifications, perhaps 100–200 times, to allow the general shape of the specimen to be seen and a particular area selected for close examination.

The electrons are obtained from a hot wire filament and then accelerated by a powerful electric field exerted by a charged metal plate to which is applied a potential of perhaps 50 kV, depending on the particular microscope design and power. This accelerating electrode has a small hole in it to allow a narrow beam of electrons to enter the main electron optical system. A system of electromagnetic lenses directs these electrons onto the object to be examined. The object, which must be thin enough to be partially transparent to the electrons, scatters the electrons from their normal path. Further electromagnetic lenses focus these electrons onto a fluorescent screen to form a visual image. This fluorescent screen may be replaced by a photographic plate if a permanent record of the image is required. It is also necessary for the complete electron microscope optics including the specimen under examination to be under high vacuum to prevent the electrons being

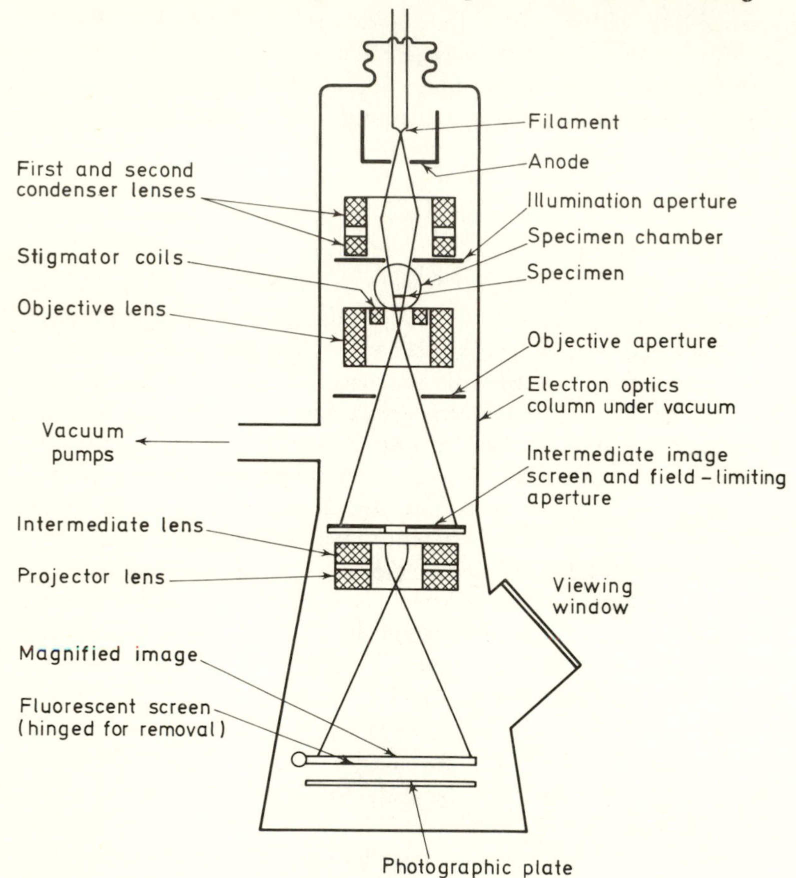

Figure 4.14. A typical arrangement of lenses in a transmission electron microscope

scattered by the gas molecules. Thus, associated with the complete electron microscope system there must be a rotary pump and diffusion pump assembly for continuous evacuation of the electron optics column. An air-lock is therefore required to enable the specimen to be inserted.

A typical arrangement of electron microscope lenses is shown in *Figure 4.14*. It is seen that the functions of the condenser and objective lenses are the same as for a normal light microscope, whilst the intermediate and projector lenses serve a similar function to the eyepiece, and the filament and anode assembly take the place of a light source. The anode is at earth potential so that the accelerating field is obtained by applying to the filament a voltage of up to about 100 kV, depending on the particular microscope and the resolution and magnification required. A small current is supplied to the filament to heat it and so cause electrons to be emitted. This current will thus control the brightness of the final image. In many designs the electron gun is a triode

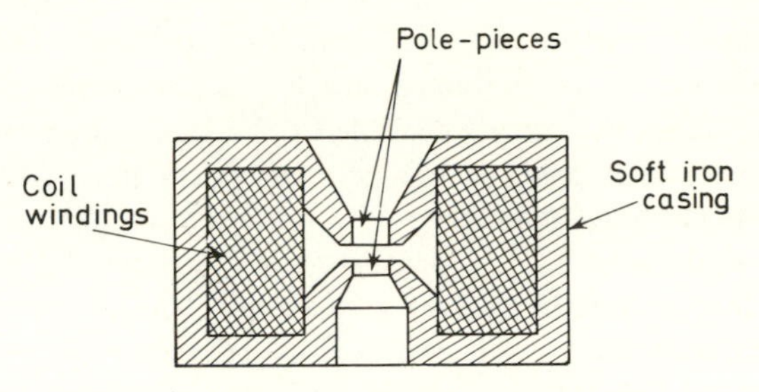

Figure 4.15. Section through a magnetic lens

arrangement in order to obtain better stabilisation of the electron beam.

The lenses themselves are usually a coil in a soft iron casing, as shown in *Figure 4.15*. As well as the main focusing coil, smaller supplementary coils, termed stigmator coils, are used to produce radial rather than axial magnetic fields. These are used to correct for astigmatism. The design of the pole-pieces allows very high magnetic fields to be obtained, giving lenses with a focal length of the order of 1 mm. To obtain these short focal lengths coupled with a sufficient freedom from aberrations, the lenses and their pole-pieces have to be very carefully designed and manufactured to very close tolerances. Various size apertures, with the smallest having a clear diameter of about 2×10^{-5} m, may be inserted by the microscope user. These serve similar functions to the apertures used in optical microscopy, such as controlling resolution and illumination, and limiting the field.

The objective lens is used to produce a magnified real image, termed the intermediate image, at the object plane of the projector lens. Electrons from a small area of this image are then focused by the projector lens to form a magnified image on the fluorescent screen. The maximum magnification produced by the projector lens is similar to that which may be produced by the objective lens but its lowest magnification is usually considerably lower than that of the corresponding objective lens. The varying magnification and focusing conditions in a particular electron microscope are obtained by altering the current through the appropriate lens coils. The image formed on the fluorescent screen may then be viewed directly or through a low-power binocular microscope. To obtain a photograph of the image, the fluorescent screen is rotated to one side to uncover the photographic film or plate.

As well as the transmission electron microscopes with magnetic focusing, microscopes have also been manufactured with electrostatic focusing systems. Although electrostatic lenses are simpler and cheaper to construct, it has not proved possible to obtain such high magnifications and resolutions as with magnetic focusing systems.

In dark-field optical microscopy the direct rays are eliminated and the image is produced by light scattered from the object. A similar technique may be used in electron microscopy. The electron beam leaving the condenser lenses is deflected by a few degrees, perhaps 3°, from its axial path and is thus prevented from passing directly into the objective lens. With an object in the beam, however, some electrons will be scattered back into the objective and transmitted through the following lenses to produce a dark-field image of the object.

It has already been said that the specimen under examination must be partially transparent to electrons and therefore it must be thinned in some way if necessary or a replica taken if the surface is to be examined. An alternative arrangement is to use a reflection system so that opaque objects may be studied. The usual method is to fit a special specimen holder and beam deflecting coils to a conventional transmission electron microscope. The specimen is mounted parallel and to one side of the axial electron beam path. Supplementary coils are used to produce a magnetic field that deflects the electron beam from its axial path and onto the specimen. Scattered electrons from the specimen are in turn deflected along the axis to pass through the objective and following lenses. The angle of incidence, as measured between the electron beam and the specimen surface, varies with the microscope design and may also be varied by the microscope user, but is generally only a few degrees, perhaps 10°, although some microscopes have been built to provide angles of incidence about two or three times this amount. Generally, however, reflection electron microscopy can only be regarded as viewing the specimen at grazing incidence. The resolution is also about two orders poorer than can be obtained with the same microscope when used for transmission microscopy.

An electron microscope may also be used to produce a diffraction pattern of a small area of the object under examination. This diffraction pattern can be used to obtain information concerning the crystal structure and orientation of the object. To select the area for diffraction examination, an aperture is inserted between the intermediate lens and the objective (*Figure 4.16*). The intermediate-lens current is adjusted to focus this aperture onto the screen, with the position of the

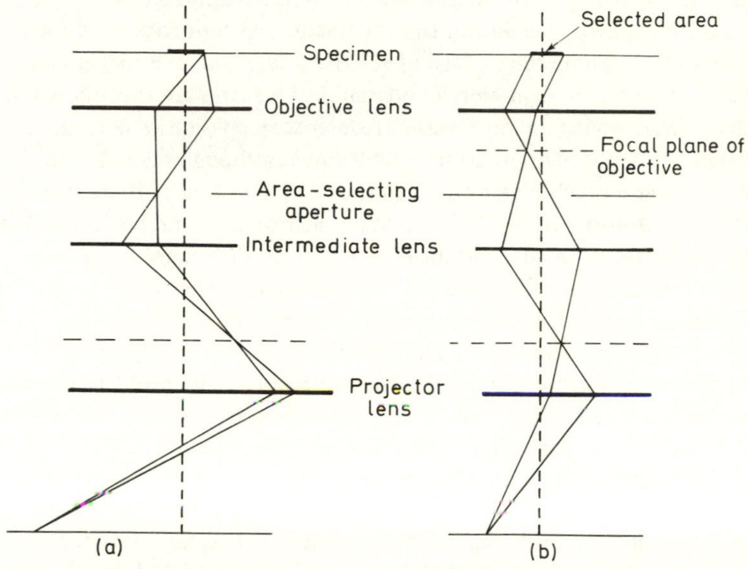

Figure 4.16. Electron ray paths through an electron microscope adjusted for (a) microscopy and (b) selected area diffraction

aperture being adjusted to select the particular area of the specimen for examination. The specimen in turn is focused on the screen by adjusting the objective-lens current. This means that the intermediate image of the specimen is now formed in the same plane as the area-selecting aperture, with the aperture isolating a small area in the plane of the specimen. The condenser should then be defocused to obtain parallel beam illumination of the specimen, as in *Figure 4.16(b)*, and the focusing of the intermediate lens adjusted to produce an image on the screen of the back focal plane of the objective. A diffraction pattern consisting of concentric rings or spots will then be seen on the screen if the specimen area under examination is at all crystalline. Concentric rings indicate a large number of very small crystals with random orientations, with the radii of the rings being a function of the lattice spacings, whilst a spotty pattern indicates relatively few, or even a single crystal, with the arrangement of spots being indicative of the crystallographic orientation.

A scanning electron microscope works on slightly different principles to the conventional transmission electron microscope and serves a different purpose. The conventional microscope is designed to give high magnification whereas the scanning version sacrifices high magnification and resolution to gain greater depth of focus and the facility to examine the surface of relatively large opaque specimens, perhaps 50 mm in diameter. Topographical features are thus shown in sharp relief, giving an impression of stereoscopic vision. The range of magnification is perhaps 20 to 50000 times, although many instruments have a much lower top magnification than this, with perhaps a typical resolution of 2×10^{-7} m. Many instruments are, however, able to resolve much smaller distances than this. In the scanning electron microscope, the first and second condenser lenses and the projector lens are used to project a much demagnified electron beam onto the specimen, as in *Figure 4.17*. The projector lens also incorporates supplementary lenses for astigmatism correction. The probe beam, with a diameter of about 10^{-8} m, is then made to scan line after line across the specimen by suitably placed scanning coils.

An image of the surface can be obtained in a number of ways and exhibited on a television screen which is scanned in synchronisation with the electron probe scan. The bombardment of the specimen surface by the high-intensity electron beam causes secondary electrons to be emitted from the specimen. If the front surface of the electron collector is positively biased, these secondary electrons will be attracted to the collector and recorded. The microscope is then said to be working in an *emissive mode*. The efficiency of production of the secondary electrons depends on the local contours of the specimen surface, which will be recorded as a varying intensity by the electron collector and displayed on the television tube in synchronisation. If the front surface of the electron collector is negatively biased, the microscope is said to be working in a *reflective mode*. With some specimens the high-intensity electron probe excites luminescence, in which case the image may be recorded by a photomultiplier tube. An alternative method of obtaining an image is to use the microscope in a *conductive mode*. With

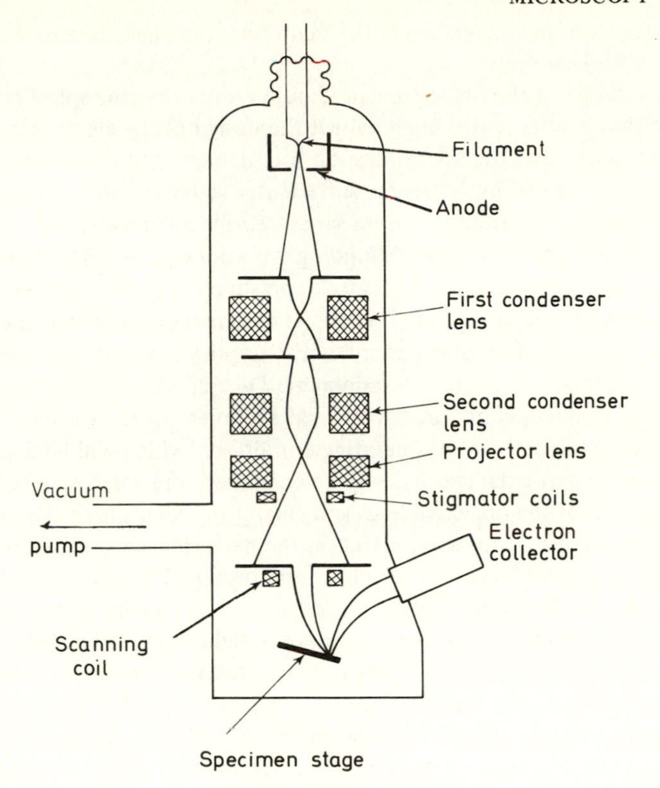

Figure 4.17. A typical arrangement of lenses in a scanning electron microscope

this, varying electron currents produced in the specimen are collected and amplified, and then supplied to the television tube in synchronisation with the scanning probe beam as before.

4.9 Setting-up Procedure

The maximum resolution obtainable with a particular electron microscope will depend on its design and the care taken in its manufacture. However, this maximum resolution cannot be obtained unless the microscope is properly set up as the designers intended and, in particular, the optics are in proper alignment and the aberrations minimised. The precise setting-up procedure will depend on the particular design and obviously the manufacturer's instructions should be followed carefully.

The design of the objective lens should ensure that the spherical aberration is sufficiently small, whilst the design of the electronics should ensure that the h.t. voltage on the filament and the current supplies to the various lenses are sufficiently stable to reduce the effects of chromatic aberration to within satisfactorily tolerable limits. Mechanical stability, owing to building vibrations, is dependent on the siting of the microscope and it is to be presumed that such an expensive instrument would not be sited where its designed performance could not be obtained. The same comment also applies to magnetic interference from other electrical equipment. The adjustments in the province of the operator are the general alignment of the various components, the choice of operating conditions, which will be dependent on the particular specimen to be examined, the correction of astigmatism of the objective lens, and finally the focusing of the image.

Apart from a detrimental effect on the performance, a misaligned optical system will make the electron microscope difficult to use. For example, the illumination may not remain centred as the intensity is varied and the image may sweep across the field as the magnification or focusing is changed, with possibly a complete loss of view of the field being examined. The tilt of the electron gun, that is, the filament and its anode, must be adjusted to send the electron beam axially down the electron-optics column. The lateral position of each of the components must also be adjusted to lie on this same axis. Thus the electron-gun assembly must be traversed with respect to the condenser lens; and the condenser lens and the electron-gun assembly must then be traversed with respect to the objective lens. Similarly, the projector lens also must be traversed with respect to the objective lens. If the system includes an intermediate lens between the objective and projector lenses, then this also must be aligned with respect to the others. Reference should be made to the instrument's instruction manual for the detailed method of alignment.

The choice of operating conditions is governed by the nature of the specimen. A typical choice of accelerating voltages might be 40, 60, 80, and 100 kV for a particular microscope. This range, however, is not large enough to give a sufficiently large change in wavelength and corresponding resolution to be of any practical significance. Therefore the choice of accelerating voltage is governed by the requirements for electron penetration of the specimen and for optimum contrast.

By varying the potential difference between the filament and anode and by adjusting the filament heating current, the electron beam current may be varied. This in turn will vary the intensity of illumination on the fluorescent viewing screen. The electron beam current, however, should be kept as low as possible consistent with satisfactory viewing conditions, to avoid heating damage to the specimen.

Apertures must also be inserted by the operator. The illumination aperture (*Figure 4.14*) is used to define the angular aperture of the illumination. The aperture is inserted in the second (or only) condenser lens and is effective at low condenser magnification. For high condenser magnifications the effective illumination aperture becomes smaller than the physical aperture but the physical aperture still serves to remove unwanted electrons that may be scattered from the main beam. The objective aperture is inserted near the back focal plane of the objective lens and serves to improve the contrast in the image by removing elastically scattered electrons. Although a smaller diameter aperture stop generally gives greater contrast, it is susceptible to the build-up of a layer of contamination on which an electrostatic charge accumulates and affects the symmetry of the electron beam. This increasing lack of symmetry must be corrected by use of the objective stigmator controls. To reduce the need for repeated astigmatic correction and frequent cleaning of the aperture stop, it is usual to use a larger aperture and reserve the use of the smaller apertures to those occasions when the nature of the specimen makes them essential and for photography.

The adjustments for the correction of astigmatism and for focus are obviously very critical, especially at high magnifications, if optimum performance is to be obtained. Adjustments are facilitated by using a thin scattering specimen in which there are some small holes, such as a thin carbon film. Interference will then occur between an electron beam passing through a hole and another beam passing through the scattering object, giving rise to Fresnel fringes around the edge of the hole. Owing to the finite size of the electron source, overlapping fringe systems are formed which blur the higher-order fringes into invisibility, leaving in general one fringe around the edge of the hole. The geometry for the formation of the fringes is such that they will disappear at focus, with the fringe spacing increasing from zero as the image becomes progressively more out of focus. To correct for astigmatism, therefore, the focus should be adjusted to produce a narrow fringe around, preferably, a circular hole. Any variation in width of the fringe around the hole will indicate astigmatism which must be compensated out by adjustment of the stigmator current controls. It is generally necessary to observe the image formed on the fluorescent screen with the aid of the supplementary binocular microscope to obtain the highest accuracy in setting. Observation of the Fresnel fringes is also an ideal method for setting the focus, as the microscope will be correctly in focus when the Fresnel fringes disappear. If there is no hole in the specimen, then a sharp discontinuity of structure may sometimes be suitable.

Many special attachments may be purchased for individual electron microscopes, for which the manufacturer's instructions for setting up and using should be consulted. These can be, for example, special stages: a tilting stage allows the specimen to be tilted through a few degrees about one or two axes, whilst a goniometer stage allows a greater angle of tilt, generally in any direction, and also allows the specimen to be rotated. Some degree of tilt is especially useful for work with metal foil specimens as small changes of tilt can sometimes produce large changes in contrast. Stereoscopic pictures may also be obtained by recording the image on film, tilting the specimen through 5° to 10°, and then photographing the image again. The two films may then be viewed through a stereo-viewer. It is, however, necessary to rotate the pictures relative to each other to compensate for the rotation introduced by the microscope lens system as the object was tilted. Hot stages are also available which are able to heat the specimen up to about 800°C; similarly, cold stages are able to cool the specimen to liquid nitrogen temperature. Liquid nitrogen cooled probes may also be inserted close to the specimen to induce the build-up of contamination on the probe rather than on the specimen. The contamination is due partly to the bombardment of the various surfaces by the electron beam, causing outgassing, but mainly to the interaction of the electron beam with organic molecules. The source of these organic molecules can be, for example, the oil in the diffusion and rotary pumps, the vacuum grease used in sealing connections, the rubber O-rings and gaskets, and even finger marks and metal surfaces that have been cleaned with organic solvents.

4.10 Specimen Preparation

Methods of specimen preparation vary considerably and will obviously depend very much on the particular specimen and the type of information required. Thus only general methods can be dealt with here since each specimen requires separate consideration. Specimens for reflection or scanning electron microscopy require little preparation in general other than polishing and etching in the case of metals. Specimens for transmission electron microscopy, however, usually require considerable preparation.

Owing to the low penetrating properties of the electron beam, it is necessary to mount specimens for transmission examination on a small grid foil. Generally these grids are of copper but other metals, such as stainless steel and platinum, are used where special conditions, for example, high melting temperatures, are required. These grids are circular with a diameter of 3 mm, although some microscopes do use a 2·3 mm diameter grid, and have various mesh shapes and sizes depending on the proposed use. Only that part of a specimen resting over a hole in the grid is available for examination. Many specimens, such as fine powder grains, need to be supported by an electron-transparent film over the grid. The two most-preferred support films are of Formvar or of carbon, although other materials such as collodion (nitrocellulose) or evaporated silicon monoxide are occasionally used.

Formvar (polyvinyl formaldehyde) films are made by dipping a microscope slide, which has been cleaned in detergent solution and polished with a cloth without rinsing in clean water, into a 0·3% w/v solution of Formvar in ethylene dichloride. The slide is removed from the solution and allowed to drain vertically onto a filter paper until dry. By lowering the slide slowly at a shallow angle into a dish of water, the film should separate from the glass by surface tension forces and float on the surface of the water. Various techniques may then be used to cut the Formvar film into small pieces and scoop them onto the grids. One method is to cut the film into small squares with a razor blade before floating off the glass slide. The ideal thickness for such a film is when it shows a pale straw interference colour in white light. A similar process is used with carbon films, which are produced by evaporating carbon onto the glass slide. A large current (e.g. 50 A at 24 V) is passed through carbon rods whilst in a vacuum system. The carbon rods should be spring loaded together and have a point contact. The arc produced at the contact is sufficient to vaporise the carbon and cause it to be vacuum evaporated. An evaporation time of about $\frac{1}{2}$ s should be sufficient to produce a film of the correct thickness, which should appear light brown in colour. The film may then be floated off in water and scooped onto the grid as before.

Having deposited the support film on the grid, the specimen to be examined must then be deposited in turn onto the support film. The technique for this will depend on the specimen. Fine powder grains may be dispersed into a cloud by a jet of air, for example, and some of the grains allowed to settle onto the grid and support film. Alternatively, it may be possible to produce a liquid suspension of the particles. A small drop of the suspension may then be placed on the grid and support film and allowed to dry, hence depositing the particles. The general problem is one of producing a satisfactory dispersion of the specimen over the supporting grid.

Bulk specimens which are opaque to the electron beam must either be thinned down or a replica taken. The method of thinning down will

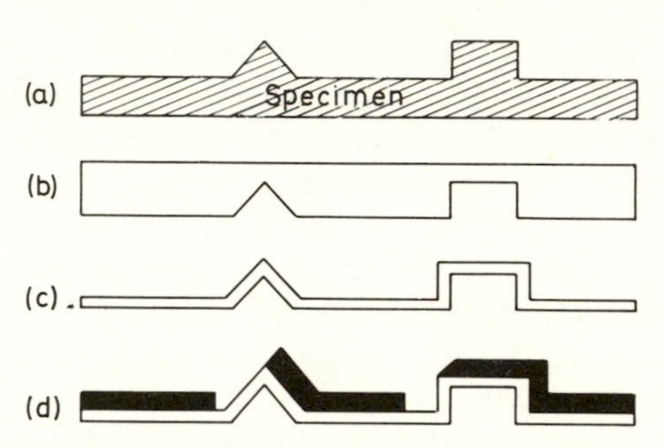

Figure 4.18. Surface replicas of a specimen (a), formed from (b) plastic, (c) an evaporated film, and (d) a shadowed replica

depend very much on the nature of the specimen; for example, biological and tissue specimens may require freezing and thin slices obtained with a microtome, whereas metal specimens may require thinning by acid etching. Many investigations, however, require a study of the surface features. These may be replicated and the replica examined under the microscope.

Two basic techniques are used in making replicas. The first involves simply depositing a replicating material such as Formvar over the specimen, and then stripping it off [*Figure 4.18 (b)*]. The second technique involves a two-stage process. In this a replica is made of the specimen and then stripped off. A further replica is then made of the first replica and it is this second replica that is examined in the electron microscope. As well as making a replica of the surface, these techniques allow small fragments to be ripped from the surface of the specimen and these fragments are then available for electron diffraction study.

Plastic replicas, however, give comparatively poor resolutions owing to their thickness. Contrast only arises owing to the small variation in thickness of a relatively thick layer. With evaporated film replicas a greater resolution can be obtained. Evaporated replicas consist of a film, for example, carbon, which is of almost constant thickness [*Figure 4.18 (c)*]. Contrast is thus produced by the changes in slope of the film, giving relatively large changes in the overall thickness. With a two-stage replica, the first could be a Formvar replica and the second an evaporated carbon replica. The plastic layer can then be dissolved away to leave the carbon replica. The contrast produced by all replicas can be increased considerably by shadowing obliquely with an electron-dense material such as a metal. This is done by evaporating the metal from a direction at a small angle to the surface. The effect is to produce 'shadows' which accentuate the surface irregularities as in *Figure 4.18 (d)*.

Five

RADIOGRAPHY

HEALTH PHYSICS

5.1 Safety Precautions

When using X-rays or radioactive sources it is most important that close attention is given at all times to ensure that any health hazards are avoided. Not only the user must be protected but also any casual passer-by. So the working area should be restricted to only those personnel who are familiar with the possible hazards and are judged competent to handle such radiation sources. Although in principle any exposure to X-rays or radioactive sources other than for medical purposes should be avoided, it is appreciated that accidents do occur and in rare cases, for example whilst rectifying a fault, it may even be necessary for a person to be exposed deliberately for a short time. However, there are government regulations which lay down maximum permissible doses for persons working with these sources which, if exceeded, may result in permanent biological effects and possibly death.

The problem of deciding what is a tolerable level of radiation is a difficult one in as much as different body cells react in different ways. Some lightly damaged cells, subjected to a small dose of radiation received over a short period, can be repaired naturally by the body, whereas even the smallest amount of radiation in the reproductive organs can lead to chromosome and gene damage resulting in mutations. There are thus varying acceptable dosages for the different organs. The problem is further complicated by the fact that different types of radiation may produce different effects on the same part of the body. A number of units have therefore been devised which try to take into account the effects produced by different radiations, as well as the differing effects on the various body organs.

A unit of radiation exposure which is based on the quantity of radiation actually absorbed is the *roentgen* (R). It is strictly applicable to only X- or γ-radiation and is equivalent to the production in 1 kg of air of ions having $2{\cdot}580 \times 10^{-4}$ C of charge of either sign. The exposure rate is usually in units of R h^{-1} or mR h^{-1}.

The unit of *absorbed dose* is the *rad*. This is equivalent to an absorption of energy of 0·01 J kg^{-1} and is independent of the material irradiated. However, these units do not take into account the different biological effects of the various types of radiation. A new unit, therefore, was introduced – the *rem*. This is the unit of *dose equivalent* and is the dose of any radiation that produces the same biological effect as one roentgen of X-rays. The rem, an abbreviation for Roentgen Equivalent Man, is the product of the absorbed dose, measured in rads, and any modifying factors such as the *quality factor* (QF) and the dose *distribution factor* (DF). Other names are also given to these modifying biological factors. For example, the quality factor is also known as the *relative biological effectiveness* (RBE) and the distribution factor as the *relative damage factor* (RDF).

The quality factor takes into account how much of the energy lost by the radiation is transferred to the body, by considering the loss of energy per unit of path travelled. This is termed the *linear energy transfer* (LET). The LET value depends not only on the type of radiation but also on its energy and obviously a greater LET value means a more biologically dangerous type of radiation. The LET value is incorporated into the quality factor by comparing the biological effect produced by the radiation with the biological effect produced by X-rays. Thus, if a dose of d rads of X-rays is required to produce the same biological effect as is produced by D rads of another radiation, this other radiation is said to have a QF of d/D. This means that X-rays have a QF of unity and that radiations with a high quality factor are relatively more damaging biologically. In fact 1 rad of alpha particles produces 20 times more biological damage than 1 rad of X-rays. However, it should be remembered that alpha and beta particles are far less penetrating than, say, X-rays or neutrons. They are, therefore, less likely to penetrate the skin and cause internal body damage. Values of the quality factor for various types of radiation are given in *Table 5.1*.

The distribution factor takes into account that there may be a non-uniform distribution of radiation over the body. Thus the DF will be unity for whole-body irradiation. It also takes into account that different body organs will be affected in varying ways, so that radiation

Table 5.1 THE RELATIVE BIOLOGICAL EFFECTIVENESS OF SOME TYPES OF RADIATION

Radiation	*QF*
X-rays	1
γ-rays	1
β-particles	1
Thermal neutrons	5
Neutrons, 0·1–10 MeV	10
Protons	10
α-particles	20

localised to a particularly vulnerable area may be just as damaging as whole-body radiation. So the dose equivalent may be expressed as

$$\text{dose equivalent (rems)} = \text{absorbed dose (rads)} \times \text{QF} \times \text{DF} \tag{5.1}$$

Maximum permissible radiation doses have been laid down for adults exposed in the course of their work. These maximum dosages, it is believed in the light of present knowledge, have a negligible probability of producing severe somatic or genetic damage. Genetic damage could result in mutations, whereas somatic effects could be, for example, skin damage, temporary or permanent sterility following irradiation of the gonads, and possibly delayed effects such as cataract of eye lenses, bone cancer, anaemia and leukaemia, as well as an increased incidence of malignancies, a shortening of the life span, and premature ageing.

In specifying the maximum permissible doses, the possible effects on four critical parts of the body are recognised. These are the blood-forming organs, the gonads, the lenses of the eyes, and the skin. For the first three, the occupational exposure is limited to a maximum of 3 rem in any period of 13 consecutive weeks, with the proviso that the accumulated dose must not exceed $5(Y - 18)$ rem where Y is the age in years of the person exposed. Thus, after the age of 18 a maximum of 5 rem per year is permissible. These maximum doses also apply for uniform irradiation of the whole body. A greater dose is, however, permitted for some less critical parts of the body, such as the hands, forearms, feet, and ankles. For these the maximum permissible dose can be as high as 40 rem in any period of 13 weeks but with the total not exceeding 75 rem in the year. The maximum permissible dose for skin is 15 rem in any period of 13 weeks, with an annual dose limitation of 30 rem.

The question of what is a permitted dose is more complicated in the case of radioactive sources that are ingested into the body by breathing or in food and drink. The sources may then come into intimate contact with critical organs, so that, for example, alpha and beta particle emitters will now be important, whereas as external sources they were of less importance owing to their low penetration. Further complications may arise as some organs have an affinity for a particular element so that, for example, the radioactive isotope strontium-90 becomes concentrated in the bones. How long an isotope may be retained by the body will depend on the particular isotope and organs concerned and is measured by the biological half-life. This is the time taken by the body to eject half of the number of originally ingested radioactive atoms. For most radioisotopes this half-life is a matter of weeks but some of the bone-seeking ones may be regarded as remaining permanently.

Doses above those laid down as being permitted, as already indicated, may lead to serious biological damage. A dose of 50 rem would probably produce slight transient blood changes but is unlikely to have any serious lasting effects. A dose of 100 rem, however, will produce fatigue and possibly vomiting, with marked changes in the blood and a shortening of life expectancy. With a dose of 400 rem, however, some deaths can be expected in possibly two weeks, with eventual short-time fatality in about 50% of exposed individuals. Adequate shielding from damaging radiation is obviously vital.

The type of shielding will, of course, depend on the type of radiation. Owing to their low penetrating powers, alpha and beta particles are readily stopped, for example, by cardboard or a thin sheet of Perspex. More penetrating radiation requires a thicker shield of high density. Thus several millimetres of lead is often sufficient to absorb X-rays whereas perhaps 2 m of high-density concrete or even 5–10 m of water are usually used for stopping high-energy neutrons.

5.2 Radiation Monitoring and Detection

When making measurements on radiation, it can generally be assumed that the type of radiation is known. Similarly, the energy distribution of the radiation is not usually required, leaving only the intensity of radiation to be measured. The three types of detector which are of greatest practical importance are the gas filled detector, the scintillation counter, and the semiconductor counter. They are all based on the property of the radiation of being able to produce some form of ionisation.

The gas filled detector generally consists of a metal cylinder in which there is an axial wire acting as an anode, as in *Figure 5.1*. The cylinder is filled with gas and, on exposing the detector to radiation, there will be some ionisation of the filler gas and electrons and positive ions will be produced. Under the applied electric field, the electrons will move towards the central wire anode and the positive ions towards the cylindrical cathode. The charge collected at the anode or cathode will

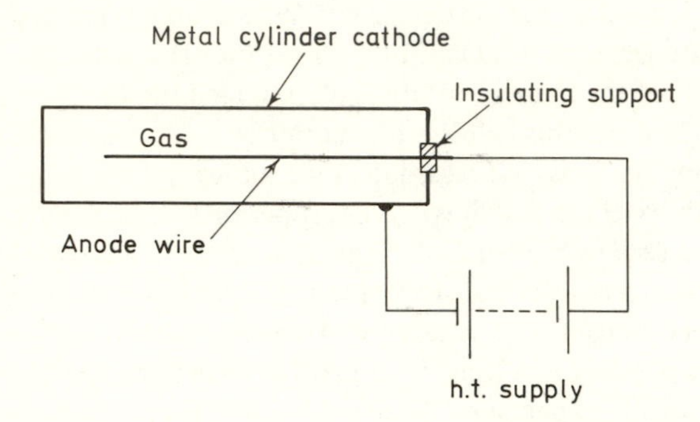

Figure 5.1. General arrangement of a gas filled radiation detector

be a function of the applied voltage and the nature and pressure of the filler gas. For low potentials the charged particles will move slowly and thus many will have time to recombine before reaching the anode or cathode. If now the applied voltage is increased, a time will come when the ions move sufficiently fast that the probability of recombination becomes negligible. Then every ion created by the radiation will be received by the anode or cathode and the collected charge will become constant for a particular radiation. The detector is then said to be working in the *ionisation-chamber region*, as shown in *Figure 5.2*. In this region the collected charge remains constant for small variations in voltage and the charge is a direct measure of the ability of the radiation to form ion pairs.

If the applied voltage is increased still further, the ions acquire sufficient energy to produce secondary ionisation so that large charges can be collected for a comparatively small initial ionisation. The total charge collected, however, will still be a definite function at a particular voltage of the initial ionisation. This region is then termed the *proportional region* and the ratio of the charge collected to the initial charge released is termed the *gas amplification factor*. This can have values up to about 10000.

Continuing to increase the applied voltage leads to a region of limited proportionality, followed by the *Geiger region*. Now the secondary emission becomes an avalanche and affects the whole of the anode wire at once. The total charge collected now becomes independent of the applied voltage and of the initial amount of ionisation. Any further increase of the applied voltage beyond the Geiger plateau region

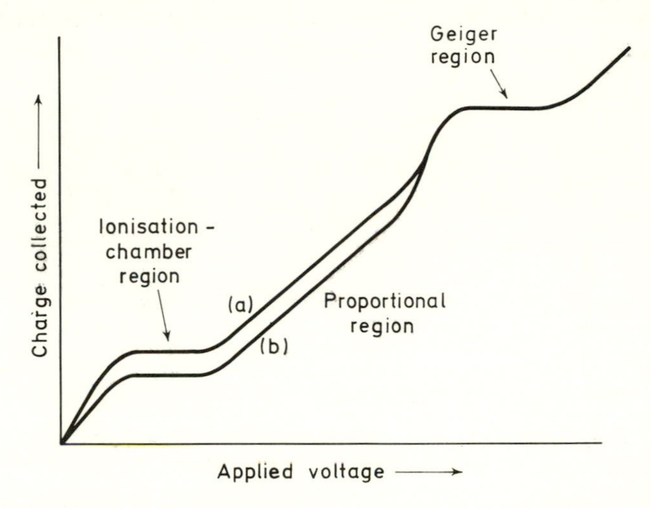

Figure 5.2. Variation of charge collected by the anode or cathode of a gas filled detector as a function of the applied voltage, due to (a) an alpha particle and (b) a beta particle

leads to a continuous discharge and is of no practical interest. In practice, radiation counters are based on the three regions indicated in *Figure 5.2*.

Ionisation chambers may be used to count separate events or the various pulses can be integrated and the chamber used as a rate-meter. Such instruments are of interest in health physics for dose rate measurement. Using argon as the filler gas, applied voltages of about 200 V and saturation currents of about 10^{-10}A are common but will depend on the nature of the filler gas, the electrode geometry, and the type of radiation being detected. The current from an ionisation chamber, therefore, requires considerable electronic amplification, whereas with a *proportional counter* much of the amplification is internal. A proportional counter does, however, require a highly stabilised power supply owing to its amplification depending critically on the applied voltage. The gas most commonly used is a mixture of argon and methane but it is most important that the mixture remains constant and free from impurities, otherwise the gas amplification factor will vary. The required applied voltage varies considerably with the geometry of the electrodes and the nature and pressure of the filler gas and may vary from perhaps a few hundred volts up to two thousand volts. A proportional counter has a maximum counting rate of about 10^6 counts per second.

A modification of the proportional counter is the gas flow type counter. This is especially suitable for use with very weak sources, such

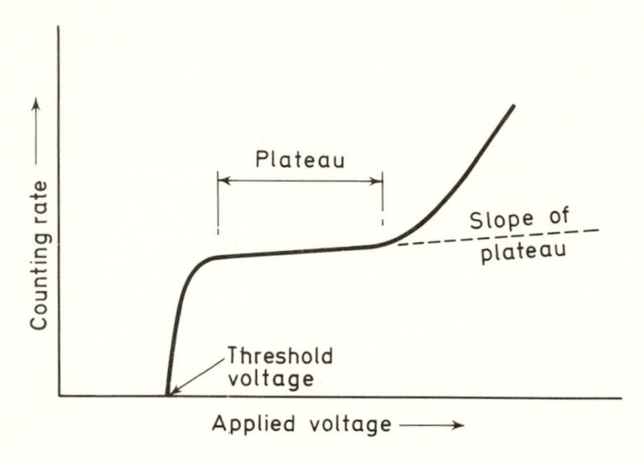

Figure 5.3. Variation of counting rate of a Geiger counter as a function of the applied voltage

as the counting of the weak beta emissions from a carbon-14 source. The flow counter is made demountable so that the source may be mounted inside it. This is particularly necessary where the radiation is of sufficiently low energy and would otherwise be unable to penetrate the counter walls. Gas is then passed continuously through the counter to maintain the correct atmosphere. By suitable electronic 'gating', a proportional counter is also able to discriminate between different radiation energies and thus different types of radiation, as in *Figure 5.2*.

The *Geiger counter* makes use of an avalanche effect in the secondary emission which makes possible gas amplification factors up to perhaps 10^8. With this amount of internal amplification, capable of producing output pulses of perhaps several volts, external electronic amplification may be unnecessary. In *Figure 5.3* it is seen that the working voltage should be adjusted to give operation over the central region of the plateau. Then small variations in the voltage have negligible effect on the counting rate. Ideally, of course, the counting rate should remain constant with voltage over the plateau region but, in practice, the plateau slopes slightly. It is, therefore, usual to quote the working voltage, the width of the plateau, and the plateau slope in specifying a Geiger counter. The sudden avalanche effect owing to the arrival of an ionising radiation may lead to the initiation of smaller spurious discharges. These may be quenched by a suitable electronic circuit but it is more usual to add a small amount of substance with a polyatomic organic molecule, such as ethyl alcohol, as the quenching agent to the argon filler gas. After a pulse the counter requires a short time, termed the *dead time*, to recover before it is ready to record another pulse. This is due to a positive space charge being left as a sheath around the wire anode, which requires time to disperse after a pulse. During the dead time any arriving ionising radiation will not be recorded. Even after the end of the dead time, the Geiger counter does not recover its full sensitivity for a short time, termed the *recovery time*. These inoperative periods limit the counting rate to a maximum of about 10^4 counts per second.

The second of the main types of counter is the *scintillation counter*. In this a *phosphor* is used as the primary detector. This has the property of converting the energy expended by the radiation into light so that the phosphor scintillates. The phosphor is also transparent to the light it emits so that a thick piece of phosphor can be used to ensure adequate absorption of the radiation and yet still transmit all the light produced. A thin film of silicone oil is used to ensure optical contact between the back of the phosphor and the front of a photomultiplier tube so that all the light produced in the phosphor can pass into the photomultiplier. When this light falls onto the photocathode of the photomultiplier tube, electrons are ejected. These electrons are accelerated towards the first electrode of the photomultiplier tube, causing it to eject secondary electrons which are in turn accelerated towards the second electrode of the series, causing it to eject more secondary electrons in its turn. This process repeats itself, producing more secondary electrons at each electrode so that with, say, 12 electrodes with a total of about 2500 V across them a gain of perhaps 10^6 can be achieved. With this type of counter very high counting rates can be achieved, as high as perhaps 10^9 counts per second but the rate will depend on the type of phosphor used.

The third main type of counter is the *semiconductor counter*. This is basically a p-n junction with a voltage across it. Radiation absorbed by the junction region produces hole–electron pairs which allow a pulse of current to pass, which may be detected in the usual way. These counters have the advantage of being small and rugged and are particularly useful for high-resolution measurements of alpha particles and protons.

As well as using counters to check the radiation levels from time to time or when some special hazard is suspected, it is now general practice for all workers who may possibly come into contact with nuclear or X-radiation to carry a film badge to record any chance exposure. This film badge is normally monitored and renewed every two weeks.

RADIOACTIVITY

5.3 The Atomic Nucleus

An atom consists of a nucleus, made up essentially from two types of particles, neutrons and protons, surrounded by an electron cloud in which the electrons are grouped in closed shells and are subject to various quantum constraints. Each proton carries a positive unit of charge, $+e$, whilst the neutron carries no charge. For electrical neutrality of an atom, the number of extranuclear orbiting electrons, each of charge $-e$, must be equal to the number of protons. This number, Z, is the *atomic number* of the atom; it determines the chemical properties of the atom and the position of that atom as an element in the Periodic Table. If the atom is ionised so that, for example, it loses an electron and becomes a positive ion, it still retains its same chemical identity. Its atomic number is unchanged, as determined by the number of protons.

The mass of an atom is almost entirely due to the mass of the protons and neutrons, which individually have almost equal masses. The unit of mass, the *unified atomic mass unit*, u, is defined as a mass equal to $\frac{1}{12}$ of the mass of a carbon atom of atomic mass number 12, that is, one containing 12 protons and neutrons. Since carbon has an atomic number of 6, its nucleus consists of 6 protons, equal to the stable number of extranuclear electrons, and, since the total number of protons and neutrons must be 12, of 6 neutrons. Then 1 u = 1·6575

$\times 10^{-27}$ kg. The mass of a proton is 1·007825 u, of a neutron 1·008665 u, and of an electron 0·0005486 u. The atomic *mass number*, in which the nuclear mass is given in terms of units of the proton mass, which is assumed to have approximately the same mass as a neutron, is thus an integral value and approximates to the nuclear mass in unified atomic mass units. To denote the mass the chemical symbol can be written, for example, as $^{23}_{11}Na$, where the 11 indicates the atomic number, the position in the Periodic Table and therefore the number of protons in the nucleus, and the 23 denotes the atomic mass number. There are thus 12 neutrons (i.e. $23 - 11 = 12$) in this nucleus. Since, however, the element in this example is already named as sodium (Na), the atomic number is superfluous and this particular nuclide could be designated as ^{23}Na. *Nuclides* are atoms with the same atomic number and also having the same number of neutrons in their nuclei as each other.

Most atoms do not have a specific number of neutrons. For example, hydrogen, atomic number 1, has one proton but can have zero, one, or two neutrons. The resulting atoms are termed isotopes of hydrogen. Thus a one-proton nucleus forms hydrogen, $^{1}_{1}H$; one proton with one neutron is deuterium, $^{2}_{1}H$; and one proton with two neutrons is tritium, $^{3}_{1}H$. Nevertheless all three isotopes behave chemically as hydrogen. Tritium is an artificially produced isotope but deuterium occurs naturally. The *abundance* of deuterium is 0·016% – that is, in a sample of naturally occurring hydrogen, 0·016% by weight is in the form of deuterium atoms. The *atomic weight* is then the weighted mean of the masses of the naturally occurring atoms, taking into account their relative abundances. The mass number is thus the nearest integer value to the atomic weight.

The mass of one isotope does not differ from another by exactly an integral multiple of the neutron mass. Thus, for example, the mass of a deuterium atom is 2·014102 u whereas the mass of a hydrogen atom plus a neutron is 2·016490 u. The difference is termed the *mass defect* and is due to energy being released when a proton and a neutron are brought together to form a deuterium nucleus. Conversely, this same amount of energy must be supplied to split a deuterium nucleus. This energy, termed the *binding energy*, is given by Einstein's equation

$$E = \Delta mc^2 \tag{5.2}$$

where Δm is the mass defect and c the velocity of light. The energies involved, as are those of other nuclear reactions, are generally measured in electron volts, eV. This is the energy acquired by an electron accelerated through a potential difference of 1 V. That is, 1 eV $\equiv 1{\cdot}6021 \times 10^{-19}$ J and 1 u $\equiv$ 931·48 MeV $\equiv 1{\cdot}49 \times 10^{-10}$ J.

5.4 Radioactivity

Most atoms have isotopes, that is, for a given number of protons (a given element), there is a variable number of neutrons. In most cases a naturally occurring isotope has a quite stable arrangement but in some cases it is necessary for the nucleus to emit energy either as a particle or as electromagnetic radiation in order to attain a stable condition. An isotope that emits energy in this way is said to be *radioactive*. There are in fact about 50 naturally occurring radioisotopes and about another 850 that are artificially produced.

Although there are a number of possible types of emission from a nucleus, the emission of an alpha or a beta particle or a gamma ray constitute the most important mechanisms. An *alpha particle* is the nucleus of a helium atom (atomic number 2) and as such consists of 2 protons and 2 neutrons. It can thus be regarded as a doubly ionised helium atom, that is, a helium atom stripped of its two extranuclear electrons. Since an alpha particle has a charge of $+2e$ and a mass of approximately 4 u, a radioisotope that emits an alpha particle will become an isotope of a new element two positions down in the Periodic Table. A typical transition would be

$$^{210}_{84}\mathrm{Po} \rightarrow {}^{206}_{82}\mathrm{Pb} + \alpha \tag{5.3}$$

That is, an atom of polonium (Po), with an atomic number of 84 and mass number of 210, spontaneously emits an alpha particle (α) by losing 4 mass units and 2 units of positive charge from its nucleus. It thus becomes a new element with a mass number of 206 and with a nuclear charge of 82 units. This is thus an element with 82 protons and is consequently element number 82 in the Periodic Table, which is lead (Pb). The alpha particle could also be written as $^{4}_{2}\mathrm{He}$. As mentioned in the previous Section, there will also be a small discrepancy between the masses on either side of the equation which will manifest itself as energy according to equation 5.2 The greater the nuclear mass, the greater is the likelihood of an alpha particle decay, with elements above atomic number 82 being the main emitters. An alpha particle produces strong ionisation and therefore its energy is easily absorbed by other atoms.

For example, the range of alpha particles in air is only a few centimetres. It is also readily deflected by an electric or magnetic field.

If now there are too many neutrons to protons in the nucleus for stability, then a *beta particle* will be emitted. In this process a neutron decays into a proton and an electron, with the electron being ejected from the nucleus. This ejected electron is termed a beta particle. Negligible change is made in the mass of the nucleus so that the mass number may be regarded as being unchanged, but the nucleus is made one unit more positive by the loss of this beta particle with its one unit of negative charge. The newly created element is thus one position higher in the Periodic Table but with an unchanged mass. A typical reaction would be

$$^{239}_{92}\mathrm{U} \rightarrow {}^{239}_{93}\mathrm{Np} + \beta^- \tag{5.4}$$

where the symbol β^- indicates a beta particle with its one unit of negative charge, as distinct from a β^+ particle which has one unit of positive charge and is a positron. A positron will be emitted by a nucleus if there are too many protons to neutrons for stability. In this case a proton may eject a positron and so decay to a neutron. A typical reaction would then be

$$^{22}_{11}\mathrm{Na} \rightarrow {}^{22}_{10}\mathrm{Ne} + \beta^+ \tag{5.5}$$

Again there is no change in mass number but this time the nucleus has lost a proton and the newly created element is thus one position down in the Periodic Table. This type of reaction is about five times less common than a beta decay process. The positron also has a very short life, of the order of 10^{-10} s, as it is annihilated on meeting an electron. This annihilation process results in the positron and electron masses each being converted into an energy of 0·51 MeV according to equation 5.2, with the energy manifesting itself in two gamma rays. Beta particles are ejected with very high velocities, which may approach that of light, 3×10^8 m s^{-1}. Being of very low mass, beta particles are very easily scattered which limits their penetration to a fraction of a millimetre of aluminium for example, depending on their energy. They are also deflected by electric and magnetic fields.

Gamma ray emission from a nucleus occurs when a nucleus in an excited state changes into a more stable state. In changing into the lower-energy state, excess energy is emitted in the form of an electromagnetic radiation of frequency ν, given by

$$E_2 - E_1 = h\nu \tag{5.6}$$

where $E_2 - E_1$ is the change in energy in the nucleus and h is Planck's constant, equal to $6{\cdot}626 \times 10^{-34}$ J s. The energy of the gamma ray is thus $h\nu$; its wavelength is less than about 10^{-10} m. Both alpha and beta particle emissions are often accompanied by gamma ray emission. Alpha and beta particles, having opposite charges, are deflected in opposite directions by an electric or magnetic field, whereas gamma rays, being an electromagnetic radiation, are unaffected by such fields. Gamma rays are also far more penetrating than alpha or beta particles.

A number of other emission processes occur, as for example the emission of neutrons and the various types of mesons. Apart from neutrons, however, which are of importance in nuclear reactor technology, these other emissions are of little practical importance. By bombarding stable isotopes with, for example, alpha particles, new radioisotopes may be created. A typical reaction would be

$$^{4}_{2}\mathrm{He} + {}^{14}_{7}\mathrm{N} \rightarrow {}^{17}_{8}\mathrm{O} + {}^{1}_{1}\mathrm{H} \tag{5.7}$$

in which an alpha particle ($^{4}_{2}\mathrm{He}$) is used to bombard a nitrogen nucleus to produce a radioisotope of oxygen, together with a proton ($^{1}_{1}\mathrm{H}$). The bombarding alpha particle must have sufficient kinetic energy to penetrate the electron cloud surrounding the nitrogen nucleus.

An important factor that concerns the practical use of radioactivity is the time over which the emissions take place. A particular radioisotope will decay spontaneously with the nucleus taking up a new state and emitting a radiation in what may be regarded as an instantaneous process. However, the instant at which a particular nucleus will decide to decay is governed by random processes so that the number of nuclei that decay in a particular time interval is a function of the number of undecayed nuclei still present.

The probability that a particular radioactive nucleus will decay in a time interval $\mathrm{d}t$ is $\lambda \mathrm{d}t$ where λ is the *radioactive decay constant.* This constant, it is assumed, is independent of time and will be the same for all nuclei of that particular radioisotope. If there is a total of N atoms of the radioisotope present at a time t, the number that will decay in the time interval between t and $t + dt$ will thus be $\lambda N\,dt$. The decrease in the number N of nuclei available for decay is then

$$dN = -\lambda N\,dt \tag{5.8}$$

where the negative sign indicates that the number of available nuclei N decreases with increasing time. Equation 5.8 can be integrated to find the number of undecayed nuclei N_t remaining after a time t, starting from the initial condition of N_0 nuclei existing at a time $t = 0$. That is,

$$\int_{N_0}^{N_t} \frac{dN}{N} = -\lambda \int_0^t dt \tag{5.9}$$

Therefore

$$\ln N_t - \ln N_0 = -\lambda t \tag{5.10}$$

i.e.

$$N_t = N_0\,e^{-\lambda t} \tag{5.11}$$

It is also seen that the total number of disintegrations N' taking place in this time t is given by

$$N' = N_0\,(1 - e^{-\lambda t}) \tag{5.12}$$

Equation 5.11 shows that the number of undecayed nuclei falls off exponentially with time at a rate determined by the decay constant. In particular, the time taken for the number of undecayed nuclei N_t to fall to $N_0/2$, that is, half the number initially present, is termed the *half-life*, T. This is simply determined from equation 5.10, since

$$\ln\frac{N_0/2}{N_0} = -\lambda T$$

i.e.

$$T = \frac{\ln 2}{\lambda} = \frac{0{\cdot}693}{\lambda} \tag{5.13}$$

A different approach to the rate-of-decay problem is to remember that after a period equal to the half-life has elapsed the number of radioactive atoms of the original type will have been reduced by a factor of a half. After a period t, the number of elapsed half-lives will be t/T, where T is the time for one half-life, so that the number of radioactive atoms originally present will have been reduced by a factor $(\frac{1}{2})^{t/T}$. That is, the number remaining after a time t will be

$$N_t = N_0(\tfrac{1}{2})^{t/T} \tag{5.14}$$

The half-life of a radioisotope is a characteristic of that particular radioisotope and cannot be changed by any known physical or chemical processes. Depending on the particular radioisotope, it can have values that vary over extreme ranges. For example, the artificially produced isotope beryllium-6 has a half-life of 4×10^{-21} seconds, whereas at the other end of the scale potassium-40 has a half-life of about $1{\cdot}3 \times 10^9$ years. This means that it will take about $1{\cdot}3 \times 10^9$ years for half of a large number of atoms of the particular radioactive isotope of potassium with a mass number of 40 to decay by some form of radioactive emission process. Since this is an exponential decay process, it obviously does not mean that all the atoms will decay in twice this period. In any practical use of radioactive isotopes, a radioisotope must be chosen with a suitable half-life to allow sufficient counts to be accumulated to compensate for the random nature of the process and obtain an accurate result. For medical uses, however, where it may be required to ingest the radioisotope into the body, it is desirable for the half-life to be sufficiently long for the purpose of the experiment but also short enough that any biological damage produced is negligible.

The quantity λN is termed the *activity* of the particular radioactive sample, whereas the *specific activity* is the activity of unit mass of that sample. The specific activity of 1 kg is then given by

$$\text{specific activity} = \frac{\lambda N_A}{A} \text{ disintegrations s}^{-1}\text{ kg}^{-1} \qquad (5.15)$$

where A is the atomic mass number and N_A is Avogadro's number, $6{\cdot}0225 \times 10^{26}$ kg-mole^{-1}. The unit of activity is the *curie*, Ci, which is defined as that quantity of radioactive nuclide in which the number of disintegrations occurring in one second is $3{\cdot}700 \times 10^{10}$. This is in fact the activity of 1 g of radium-226. Therefore the specific activity can be expressed as:

$$\text{specific activity} = \frac{\lambda N_A}{A \times 3{\cdot}7 \times 10^{10}}$$

$$= 1{\cdot}628 \times 10^{16} \frac{\lambda}{A} \qquad (5.16)$$

$$= \frac{1{\cdot}128 \times 10^{16}}{AT} \text{ Ci kg}^{-1} \qquad (5.17)$$

where A is the atomic mass number, T the half-life, and λ the decay constant of the particular sample of radioisotope.

5.5 *Uses of Radioactive Sources*

The majority of uses of radioisotopes arise from the penetrating properties of their emissions, although it should be remembered that, the greater the penetrating effect of the radiation, the greater is the difficulty in containing the radiation and shielding the operator from biological hazard. Alpha particles, beta particles, and gamma rays have progressively greater penetrating properties. The source should also have a sufficient activity to ease the problems of counting. A generally

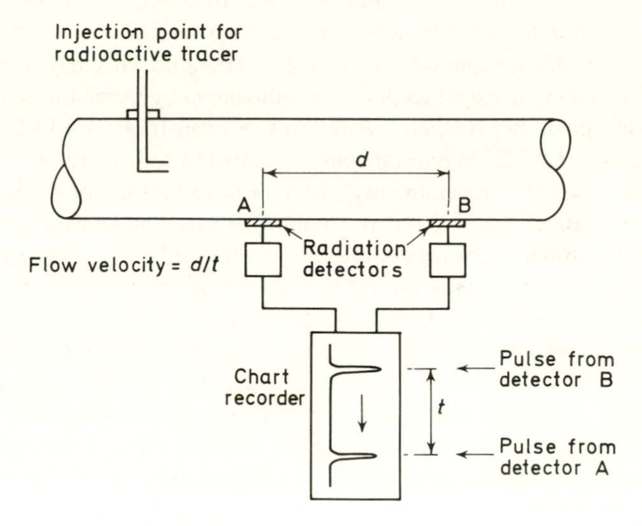

Figure 5.4. A method for using a radioactive tracer for the determination of the velocity of a flowing liquid

satisfactory rate of counting is about 4 disintegrations per second and, since 1 Ci corresponds to $3{\cdot}7 \times 10^{10}$ disintegrations per second, a source of approximately 10^{-10} Ci is usually most suitable. This fact in conjunction with equation 5.17 allows a suitable quantity of the radioisotope to be chosen.

Of the many practical uses of radioactivity, only a few applications can be given here but these should be indicative of other possible applications. *Figure 5.4* shows schematically an arrangement for measuring the flow rate of a liquid through a pipe. A small quantity of a radioisotope, such as cobalt-60, is injected into the stream. As it passes the two radiation detectors in turn, a pulse is recorded on a chart recorder by each event. Knowing the distance *d* apart of the detectors and the time *t* between the pulses as determined from the chart recorder trace, the velocity of the flowing liquid and hence the volume flow rate may be calculated. The accuracy of this method is dependent on a rapid injection of the radioisotope for only a short period. The method works best with a turbulently flowing liquid as this ensures adequate mixing, whereas with streamline flow the radioactive tracer becomes too drawn out.

By making a moving mechanical component radioactive, any particles removed by friction and wear may be counted and hence the rate of wear determined by a test carried out over a short period. A common problem is the wear of a steel component. This may be dealt with by making the steel component radioactive by bombarding it with neutrons, possibly for several weeks, in a reactor. Iron-58, which has an abundance of 0·33% in natural iron, is converted to iron-59, which is radioactive and emits gamma rays; iron-59 has a half-life of 45 days. The wear rate of this radioactive component can then be determined under its normal operating conditions, whether it be a piston ring in an engine or a lathe cutting tool, for various periods by monitoring the activity of the lubricant. One disadvantage with this technique is that the whole component is radioactive whereas only the surface is of interest. The radioactive source is thus larger than necessary and as such may raise problems of adequate shielding. An alternative method is to electroplate the component with a radioactive metal. For example, it may be more practical to plate, say, a piston ring with chromium-51, which again emits gamma rays.

Thickness gauging is a very important use of radioactivity. This is a very simple method and only involves a radioactive source and a detector. Any material passed between these will reduce the count rate recorded by the detector and the system may then be calibrated for thickness of the sample against the count rate. Obviously a source with a long half-life is required if the calibration is to remain reasonably constant. Suitable sources are beta emitters for relatively thin or low-density materials such as paper, cardboard, or thin metal sheets and gamma ray sources for thicker samples. The method has the advantage that no contact is needed with the sample and it is thus especially suitable for thickness gauging of, for example, sheet metal whilst still being rolled. The method also lends itself to incorporation in feed-back control systems so that, for example, the separation of the rollers can be adjusted automatically to maintain the correct thickness of rolled sheet. A variation of this technique is to use back scattering. With this, the source and detector are on the same side of the sample. The geometry is such that no radiation may pass directly from the source to the detector so that any radiation detected must have been scattered back from the sample. This method is sensitive to variations in surface coatings and is applicable, for example, to the measurement of paint thickness or printing ink on rollers. Many different sources are suitable for thickness gauging but the final choice is determined by the actual thickness and density of the sample and the sensitivity required. For example, a strontium-90 gamma source is suitable for steel plate up to about 0·5 mm thick. Since it is really the mass of the sample between the source and detector which is of importance, a useful measure of 'thickness' that has been commonly used is a unit of one milligramme per square centimetre (mg cm^{-2}, where 1 mg $cm^{-2} \equiv 10^{-2}$ kg m^{-2}). This is the mass of sample in unit cross-section of the radioactive 'beam'. For example, cobalt-60, a gamma source with a half-life of $5\frac{1}{4}$ years, is useful for thickness measurements over the range 20–2000 kg m^{-2}, whilst a strontium-90 beta source with a half-life of 25 years is useful up to a thickness of 6 kg m^{-2}.

The great penetrating properties of gamma rays may also be utilised for radiography. A gamma ray source is placed to one side of the object to be radiographed and a photographic film on the other side. The less dense parts of the object allow greater transmission of the gamma rays with a darkening of the corresponding area of the photographic film. A commonly used source for gamma radiography is cobalt-60, which is suitable for steel specimens up to about 0·25 m thick. The technique has the advantage that the equipment costs are low and the equipment itself is moderately easy to transport. It is therefore very useful for operation in remote locations or confined spaces. It does, however, have one disadvantage compared with using X-rays in that a gamma source cannot be switched off and consequently be made safe.

In autoradiography, the object or part of the object is itself radioactive so that, if it is laid on a photographic film, it will produce a radiogram of itself. For example, the location of strontium-90 taken up by bones can be easily studied by autoradiography. To obtain good resolution with radiographic techniques, the film which should be of fine grain must be in close contact with the object. Also, the lower the energy compatible with reasonable exposure time, the better is the resolution.

The techniques so far described make use of the penetrating effects of the radiation but there are other techniques that make use of its ionising properties. Such radiation can be used for killing bacteria and thus used for sterilising food or surgical dressings. However, fairly large doses are required, up to about 10^6 rad, which may produce objectionable flavour changes with some foods.

By ionising the atmosphere in the region of static electricity build-up, the static may be leaked away. This is especially useful in the weaving and printing industries where cloth and paper moving at high speed through processing machines can lead to their surfaces becoming electrically charged. This charge attracts dust which can mar the print and, in particular, can lead to difficult handling of paper products. Alpha and beta particles are particularly suitable for eliminating static electricity as the radiation can be easily shielded.

Other uses include polymerisation in the chemical industry. If a polymer is irradiated, this can lead to a dissociation of a covalent bond and thus induce a chemical reaction. For example, large doses can produce crosslinking, or vulcanisation, of rubber.

Finally must be mentioned the use of carbon-14 for dating over the approximate range 1000–20000 years. The upper atmosphere of the earth is under continuous bombardment by cosmic rays which leads to a copious supply of neutrons being released by nuclear interactions. These neutrons then interact in turn with the components of the atmosphere. In particular they can lead to the production of carbon-14 from nitrogen-14 by the reaction

$$^{14}_{7}\mathrm{N} + ^{1}_{0}\mathrm{n} \rightarrow ^{14}_{6}\mathrm{C} + ^{1}_{1}\mathrm{H} \qquad (5.18)$$

where $^{1}_{0}\mathrm{n}$ signifies a neutron and $^{1}_{1}\mathrm{H}$ a proton. The radioisotope of carbon then decays with a half-life of 5568 ± 30 years by the reaction

$$^{14}_{6}\mathrm{C} \rightarrow ^{14}_{7}\mathrm{N} + \beta^- \qquad (5.19)$$

into nitrogen-14 again with the release of a beta particle. Most of this carbon-14 is produced in the upper atmosphere where it is incorporated into atmospheric carbon dioxide. This in turn will be taken up by growing plants and animals. It must be assumed that the intensity of cosmic rays, and with it the production of carbon-14, has remained constant in the past and that an equilibrium was soon reached in which the production of carbon-14 was balanced by its natural rate of decay. The dating works on the principle that when a plant or animal died it ceased taking up fresh carbon-14 and the carbon-14 already present continued to decay. By measuring the activity of the specimen, the period for which the carbon-14 has been decaying, which is the time since the specimen died, can be calculated. The present-day specific activity of a fresh sample of carbon-14 is $(1{\cdot}53 \pm 0{\cdot}01) \times 10^4$ disintegrations $\text{min}^{-1}\ \text{kg}^{-1}$; this is assumed to be also the initial specific activity of the specimen, that is, the activity when the specimen ceased taking up fresh carbon-14. From equation 5.11, the specific activity after time t for the ancient specimen is

$$I = -\frac{dN}{dt} = \lambda N_0\, e^{-\lambda t} \tag{5.20}$$

where λN_0 is the initial specific activity. Hence, by measuring the present specific activity I of an ancient specimen, its age t in years may be determined. For this and similar uses in which the count rate is fairly low, it is important that a background count be made. This background activity is due to radioactive particles in the atmosphere and to those naturally occurring in some rocks, as well as to cosmic rays. A separate count must be made of this constant background activity and its count rate deducted from that obtained with the specimen.

X-RAYS

5.6 The Nature of X-rays

X-rays are part of the general spectrum of electromagnetic radiation with a wavelength of the order of 10^{-10} m. They fall, therefore, between ultraviolet radiation and gamma rays in the spectrum. It should also be noted that other units have been used for measuring the wavelength of X-rays, for example, apart from the angstrom (Å), where 1 Å = 10^{-10} m, the kX unit (XU) was commonly used at one time, where 1 kX = $1{\cdot}00202 \times 10^{-10}$ m. This figure arose from early work on the measurement of wavelength in terms of the lattice spacings of calcite.

Whenever a rapidly moving electrically charged particle, usually an electron, is decelerated by striking a solid target, some of its kinetic energy, perhaps 1%, is converted into X-radiation. Those particles that lose their kinetic energy slowly, or have insufficient kinetic energy initially to cause radiation, serve only to heat the solid target, as does the kinetic energy not converted into radiation. The few electrons that lose all their kinetic energy in a single collision will produce X-radiation of the shortest wavelength, that is, with the highest value of frequency, ν_{max}, where the kinetic energy of the electron is

$$\text{kinetic energy} = h\nu_{max} \quad (5.21)$$

where h is Planck's constant, $6{\cdot}626 \times 10^{-34}$ J s. The kinetic energy of

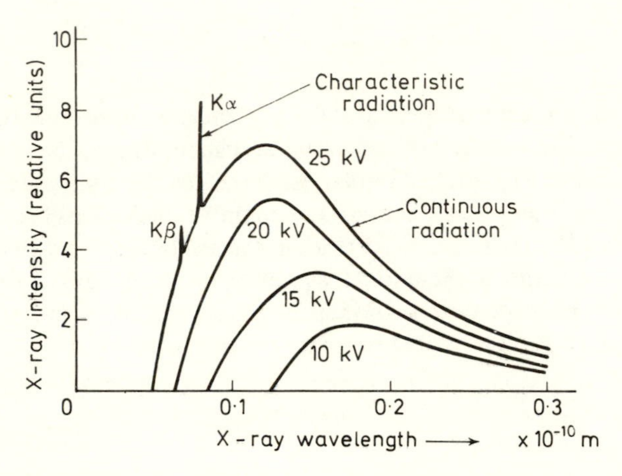

Figure 5.5. Variation of X-radiation from a molybdenum target as a function of the applied voltage

an electron of charge $e = 1{\cdot}602 \times 10^{-19}$ C accelerated through a potential difference V is eV. Therefore the X-rays produced will have a minimum wavelength λ_{min} given by

$$\lambda_{min} = \frac{c}{\nu_{max}} = \frac{hc}{eV} \quad (5.22)$$

$$= \frac{6{\cdot}626 \times 10^{-34} \times 2{\cdot}998 \times 10^{8}}{1{\cdot}602 \times 10^{-19} \times V}\text{m}$$

$$= \frac{1{\cdot}240 \times 10^{-6}}{V}\text{m} \quad (5.23)$$

where c is the velocity of electromagnetic radiation *in vacuo*, $2{\cdot}9979 \times 10^{8}$ m s^{-1}, and V is the accelerating potential in volts. For example, 50 kV would produce X-rays with a wavelength of $2{\cdot}480 \times 10^{-11}$ m or greater. *Figure 5.5* shows the variation of wavelength against intensity (relative units) for X-rays produced by electrons against a molybdenum target. Wavelengths greater than the minimum value are due to those electrons that do not give up all their kinetic energy at one collision but are scattered by the target atoms and give up a fraction of their kinetic energy at each impact. *Continuous* or *white radiation* (by analogy with white light) is thus produced, with all wavelengths being present above the short-wavelength limit as determined by equation 5.23.

The total X-ray energy emitted per second is found to be proportional to the electron current, which is obviously a function of the number of electrons striking the target in unit time, and to the atomic number of the target material, since a heavier atom will give a greater probability of an electron giving up all its kinetic energy in a relatively small number of collisions. The X-ray energy emitted is also approximately proportional to the square of the applied voltage. Thus, to produce a high intensity of white radiation, a high atomic number target metal, such as tungsten, should be used in conjunction with as high an applied voltage as is practicable.

It is also seen from *Figure 5.5* that line spectra, termed *characteristic radiation*, appear superimposed on the continuous radiation if the applied voltage is sufficiently high. The radiation is termed 'characteristic' since its wavelengths are characteristic of the target element. These emissions arise from the bombarding electrons exciting electrons from the inner shells of the target atoms into higher energy levels. As the excited atoms return to their normal energy states by allowing the outer electrons to return to vacancies in the inner shells, energy is emitted in the form of radiation, as in *Figure 5.6*. It is more probable that a K-shell vacancy will be filled by an electron from the L-shell rather than from the M-shell, so that an L- to K-shell transition leads to a more intense line, the $K\alpha$ line, than an M to K transition, the $K\beta$ line. Owing to the fine structure of the energy levels, the lines also have a fine structure so that, for example, the $K\alpha$ line is in fact a doublet, the $K\alpha_1$ and $K\alpha_2$ lines, with the $K\alpha_1$ being about twice as intense as the $K\alpha_2$. The wavelengths are very close to each other; for example, the $K\alpha_1$ line for copper has a wavelength of $1{\cdot}5405 \times 10^{-10}$ m whereas the wavelength of the $K\alpha_2$ line is $1{\cdot}5443 \times 10^{-10}$ m. There is also a $K\alpha_3$ line but this is very weak and rarely resolvable. Similarly, the $K\beta$ line is a multiplet, with the $K\beta$ radiation having about $\frac{1}{6}$ the intensity of the $K\alpha$ radiation. Vacancies are also produced in higher-order shells giving, for example, the L and M lines by transitions from outer shells. These transitions become progressively less probable so that the K lines are far more intense than the L lines, which in turn are far more intense than the M lines. The K lines are, therefore, the ones of greatest practical importance.

To produce a characteristic radiation, the bombarding electrons must have sufficient energy to excite the atom into a higher-energy state. Thus there is a critical applied voltage below which the bombarding electrons would have insufficient kinetic energy to produce excitation.

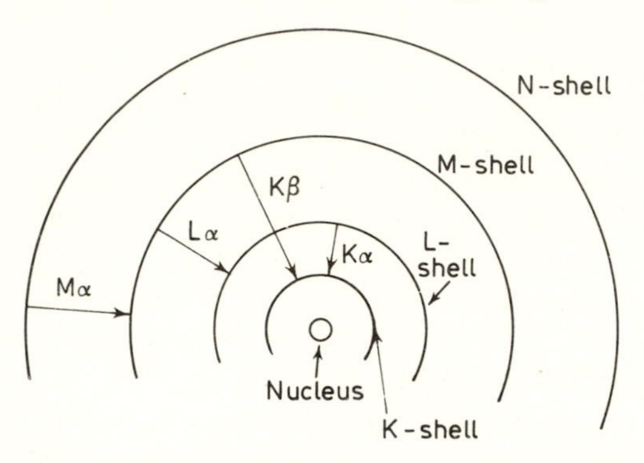

Figure 5.6. The electron transitions giving rise to characteristic X-ray spectra

However, less energy is required to excite an electron from an L-shell than a K-shell for example and, therefore, if the applied voltage is sufficient to excite the K spectrum, then the L, M, etc., spectra must also be excited.

5.7 *The Production of X-rays*

An X-ray tube comprises a source of electrons, a hot filament, with a voltage applied between the filament and a metal target to accelerate the electrons towards the target. The whole has also to be under vacuum to prevent scattering of the electrons and the filament burning out. Owing to the heating effect of the bombarding electrons, efficient cooling of the target is required. The usual system is to have a thin wafer of the target metal, which determines the characteristic radiation produced, soldered to a copper block which is water cooled. Because of the insulation problems introduced, the target, which is also the anode, is normally at earth potential and the cathode, the hot filament, is at a potential perhaps 50 kV negative with respect to earth. A typical design of X-ray tube is shown in *Figure 5.7*. As well as sealed-off tubes, such as the one shown in the figure, continuously pumped X-ray tubes

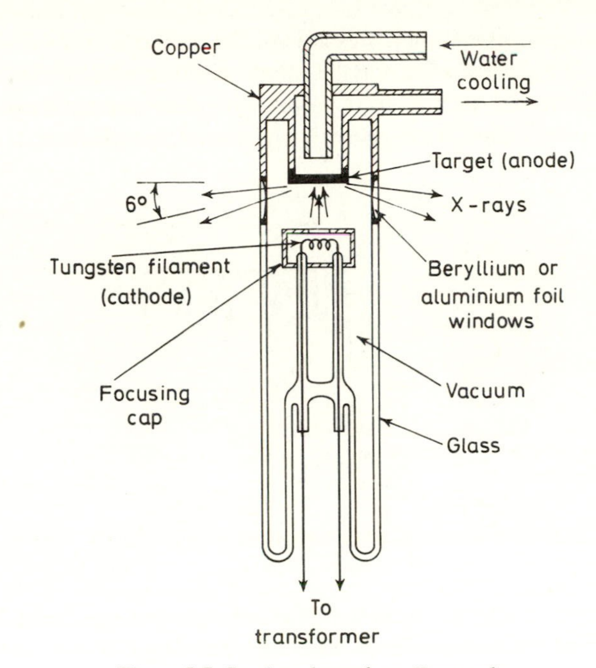

Figure 5.7. Section through an X-ray tube

are available. These have the advantage that they are demountable, enabling the target metal or filament to be changed, but have the disadvantage of requiring a subsidiary vacuum pumping system. Should it be required that the cathode be at earth potential and the anode positive with respect to earth, then oil cooling must be used.

The fact that an anode and cathode are needed does not necessarily mean that a d.c. supply is required. The X-ray tube in fact acts as its own rectifier with the electrons being accelerated towards the target during only that part of the cycle in which the filament is negative with respect to the target. A suitable electrical circuit is shown in *Figure 5.8*. The milliammeter measures the flow of electrons from the filament to the anode, the beam current, which is typically between 15 mA and 40 mA. This is controlled by adjusting the filament voltage, of the order of 5 V, which in turn controls the filament current and consequently its temperature and the number of electrons emitted. However, the power supply to the filament is nominally at the same voltage as the cathode and, therefore, the filament transformer must be well insulated from earth if the conventional earthed-anode arrangement is used. Although the voltmeter measures the voltage on the primary of the high-voltage transformer, it may be suitably calibrated in kilovolts to indicate the voltage applied to the X-ray tube. This voltage is varied by adjusting the voltage on the primary with an autotransformer. With this

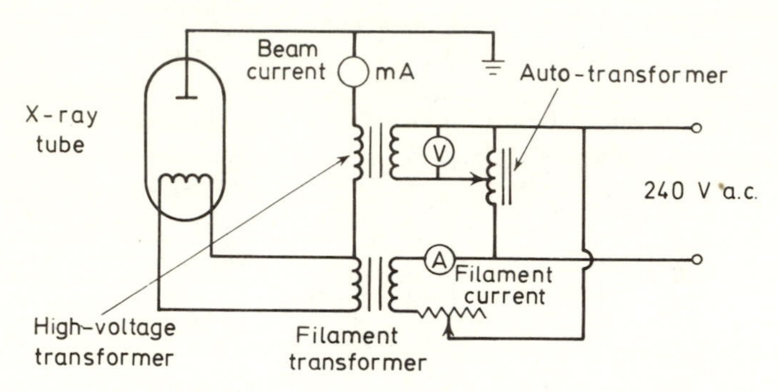

Figure 5.8. Electrical circuit for a self-rectifying X-ray tube

circuit arrangement, the voltmeter does not have to be at the high operating potential of the X-ray tube, thus avoiding difficult insulation problems.

Focusing of the electrons onto the target can be achieved by a suitably positioned metal cup surrounding the filament. With some designs, this focusing cup is at the same potential as the filament but in others it is made a few hundred volts negative with respect to the filament, which enables a very small focal spot to be achieved. The filament is normally in the form of a helical coil so that a rectangular focal area is formed on the target. The windows around the X-ray tube are usually arranged so as to give either a side or an end view to this area. The X-ray intensity is greater with a take-off angle of about 6° so that the windows will give much foreshortened views of the rectangular focal area. Thus from two windows a line source will be obtained and from the other two a point source. The line source is of use with some designs of diffractometer, whereas the point source is of more general use. This 'point' source is in fact more usually about a 1 mm square source, although it may be considerably less in specially designed fine-focus tubes. Since the production of X-rays from an X-ray tube is only about 1% efficient, the maximum power that can be obtained is limited by the heat that can be dissipated from the target. Thus the output from a fine-focus tube must be kept comparatively low, otherwise the intense electron beam will melt a hole in the target.

Where a more constant emission of X-rays is required, the circuit of *Figure 5.8* is not suitable. It is then more usual to use a voltage-stabilised d.c. supply, which requires high-voltage rectifier tubes and smoothing capacitors and ancillary electronics. When using high-voltage rectifier tubes, care must be taken to ensure adequate shielding as the rectifier tubes themselves act as producers of X-rays. They are often mounted within the high-voltage transformer so that the transformer can provide the shielding.

5.8 *Absorption of X-rays*

When X-radiation passes through matter, some of the radiation is absorbed. Some of it will be absorbed by simply being scattered from the direct beam by atoms and absorbed in the sense that energy is lost from the beam. Most absorption will occur, however, owing to transitions produced in atoms. In the same way that electrons of sufficient energy can interact with atoms to produce an electron vacancy in, say, the K-shell, which on being filled by an outer electron causes a characteristic K radiation to occur, so also can X-radiation of sufficient energy produce the same effect. Now, however, the ejected electron is called a *photoelectron* and the emitted characteristic radiation *fluorescent radiation*. This fluorescent radiation is of the same wavelength as the characteristic radiation of that element. Thus, if continuous X-radiation passes through, say, a sheet of copper, some radiation of all wavelengths will be absorbed by scattering losses. With decreasing wavelength, however, there will occur a marked increase in absorption at a particular wavelength, termed the K absorption edge, which for copper occurs at a wavelength of $1{\cdot}3804 \times 10^{-10}$ m. At this wavelength λ_K and corresponding frequency ν_K, the energy per quantum of the radiation is given by $h\nu_K$ such that

$$E_K = h\nu_K = \frac{hc}{\lambda_K} \qquad (5.24)$$

where h is Planck's constant and c the velocity of electromagnetic radiation *in vacuo*. This is then the right energy to eject an electron, a photoelectron, from the K-shell, hence giving rise to fluorescent radiation. Since energy must be conserved in the process, the incident quantum of energy must be equal to the energy of the resulting quantum of fluorescent radiation together with the kinetic energy of the ejected photoelectron. Thus the wavelength λ_K of the K absorption edge must be shorter than that of any K characteristic radiation. For lower wavelengths of the incident radiation, there is sufficient energy to produce ejection of photoelectrons and fluorescent radiation but now the incident radiation is becoming sufficiently energetic to pass right through the absorbing medium with very little interaction with the atoms. There is thus less absorption of the shorter wavelengths. If, now, less-energetic incident radiation is considered, that is, of longer wavelength than, say, that of the K absorption edge, then similar processes will apply to give L, M, etc., absorption edges. In fact there are three L edges, five M edges, etc., arising from the multiplicity of the possible transitions between the atomic shells.

With monochromatic radiation, the intensity I_0 of an incident radiation is reduced to an intensity I after passing through a thickness t of absorbing medium, where

$$I = I_0 \, e^{-\mu t} \tag{5.25}$$

where μ is termed the *linear absorption coefficient*. This is more conveniently rewritten – to take into account that the mass of absorbing material is of importance – as

$$I = I_0 \, e^{-(\mu/\rho) \rho t} \tag{5.26}$$

where μ/ρ is termed the *mass absorption coefficient* and ρt is the mass per unit area of the absorbing layer, where ρ is its density. The mass absorption coefficient is a constant of the absorbing material and is independent of its physical state. For a mixture of elements, its value is taken as the weighted mean of the individual coefficients. It is also found that between adjacent absorption edges the variation of mass absorption coefficients with wavelength λ is of the form

$$\frac{\mu}{\rho} = kZ^4 \lambda^3 \tag{5.27}$$

where Z is the atomic number of the absorbing element. The power to which the wavelength is raised is not exactly 3 but in fact varies, mainly between $2\frac{1}{2}$ and 3, and the value of the constant k is different for each region between the absorption edges. However, from equation 5.27 it is seen that there is very high absorption of the long-wavelength X-rays and by elements high in the Periodic Table. The easily absorbed long-wavelength X-rays are said to be *soft*, whereas the highly penetrating ones are said to be *hard*.

Although K, L, etc., absorption edges may exist in theory for all elements, the longer-wavelength X-rays are so easily absorbed that only the short-wavelength edges are generally detectable. For example, the characteristic K radiation of sodium (atomic number 11), of wavelength $11{\cdot}909 \times 10^{-10}$ m, is heavily absorbed by air. K edges and characteristic K radiations are only detectable in practice for elements above about sodium in the Periodic Table. Similarly, L edges and radiations are only detectable for elements in the second half of the Table and M edges and radiations for the very heavy elements, since it is only with the higher elements that short wavelengths are involved. However, for the very heavy elements the K edges and characteristic radiations have extremely short wavelengths, for example the mean $K\alpha$ radiation for uranium (atomic number 92) has a wavelength of $0{\cdot}128 \times 10^{-10}$ m, and they become very difficult to excite and to measure. The variation of the absorption coefficient with wavelength is shown for molybdenum (atomic number 42) in *Figure 5.9*.

The property of absorption edges can be utilised to absorb selectively a $K\beta$ radiation to leave a monochromatic $K\alpha$ radiation. For example, the K absorption edge of nickel occurs at a wavelength of $1{\cdot}488 \times 10^{-10}$ m, whereas the $K\beta$ and mean of the $K\alpha$ radiations for copper are $1{\cdot}392 \times 10^{-10}$ m and $1{\cdot}542 \times 10^{-10}$ m respectively. Thus the copper $K\beta$ radiation, being on the short-wavelength side of the absorption edge of the nickel, is heavily absorbed. However, the $K\alpha$ emission is also absorbed but, for example, if the nickel absorption-filter is of a thickness to reduce the $K\alpha$ radiation to half its original intensity, the $K\beta$ radiation will be reduced from about 1/6 of the $K\alpha$ in the unfiltered radiation to about 1/500 with the filter.

5.9 X-radiography

Most uses of X-rays utilise the fact that their wavelengths are of the same order as the lattice spacings of crystalline solids. The characteristic

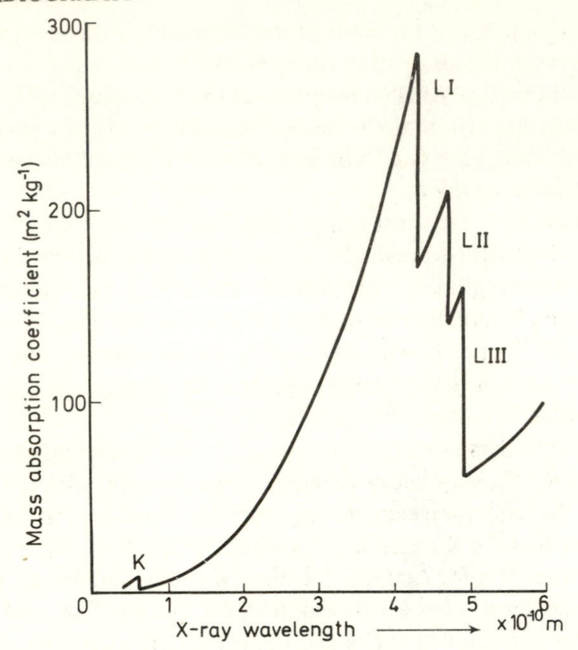

Figure 5.9. Variation of the mass absorption coefficient of molybdenum with wavelength, showing K and L absorption edges

radiations of X-rays, therefore, are used in both diffraction and spectroscopy, which are discussed in the next two chapters. Radiography, however, makes use of the penetrating properties of X-rays.

An X-ray tube used for radiography is normally of the sealed-off variety rather than a demountable one whose vacuum is maintained by continuous pumping. A sealed-off tube complete with its power cable and water cooling pipes is the more easily positioned for radiographing awkward objects. On the side of the object remote from the X-ray tube is placed a film in a lead-backed cassette. The distance of the object from the X-ray tube is governed by the divergence of the beam and the size of the object but perhaps a metre or more away is typical. Depending on the density and thickness of the object and the sensitivity of the recording film to X-rays, a sufficient exposure time is then given. Less-absorbing areas will then show up darker on the film owing to the relatively greater exposure. This method is suitable for detecting small flaws such as hairline cracks and microporosity. For coarser flaws in, say, small light alloy castings, a fluorescent screen can be used in place of the photographic film. This screen is then viewed using a mirror at 45° to protect the operator from the direct beam. Even so the operator should be behind a lead-glass screen to protect him from scattered radiation. Using a long pair of tongs and protective gloves of leaded-rubber, the operator can manipulate the object for best viewing of the flaws. The radiography of welds, for example, may require viewing from several directions; with experience, various types of flaws may be recognised and their position within the object determined.

The definition of the radiograph depends on a number of factors. Thus, grain size of the photographic emulsion is of importance. A slow,

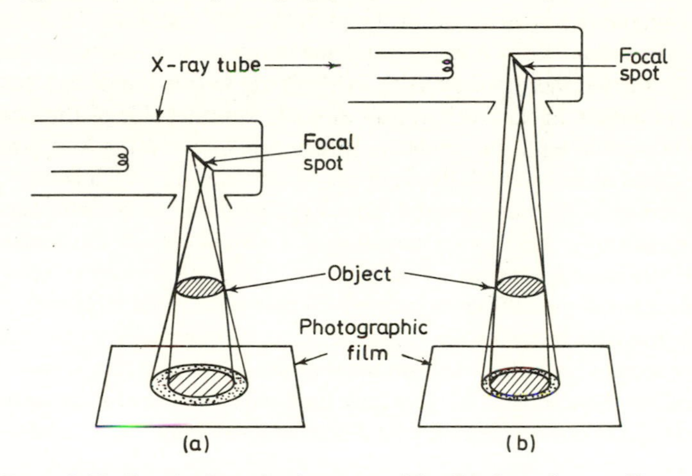

Figure 5.10. Showing how the sharpness of detail is dependent on distance

high-contrast, emulsion has a small grain size, whereas a fast emulsion has a larger grain size. A compromise choice must, therefore, be made between speed and resolution. The focal spot of the X-ray tube has a finite size which will lead to a lack of sharpness in detail, as in *Figure 5.10 (a)*, but this situation can be improved upon by increasing the distance to the X-ray tube, as in *Figure 5.10 (b)*. However, it should be remembered that increasing the distance of the object to the focal spot of the X-ray tube causes the intensity of radiation to fall off inversely as the square of this distance, with a corresponding increase in exposure time. Other factors that affect the definition are the distance from the film of a flaw within the object and also the scattering of the X-ray beam within the object. Internal scattering is especially troublesome since the scattered radiation is of longer wavelength than the primary radiation and therefore has a greater effect on the photographic emulsion, giving an overall increase in the background level of exposure and thus reducing contrast. This long-wavelength radiation can be reduced somewhat by using a metal sheet between the object and the photographic film to absorb preferentially the long wavelength and consequently less penetrating radiation. Suitable metal filters are brass or copper sheet with most light alloy objects and thin lead for use with steel objects.

Photographic emulsion is not, however, very sensitive to X-radiation, especially the very short wavelengths, as it passes right through with little interaction with atoms of the photographic emulsion. To increase the interaction, X-ray film has emulsion layers on both sides. Interaction can be further increased by using *intensifying screens*; these can be thin foils of a dense metal, such as lead or gold, on either side of the film and in contact with the emulsion. Interaction between the X-rays and the atoms of the metal foils produces photoelectrons which cause blackening of the photographic film. Even better as an intensifying screen is a layer next to the emulsion of a material which fluoresces under the action of X-rays. For example, a layer of calcium tungstate emits a bluish-green fluorescence which may reduce the required exposure time by perhaps a factor of 50 to 100 times. However, intensifying screens have the disadvantage that the contrast is also reduced and therefore these screens are generally used only with large steel castings where the exposure time may be inconveniently long.

Six

DIFFRACTION

OPTICAL DIFFRACTION

6.1 Diffraction

According to Huygens' principle, each point on a wave front may be considered as a new source of secondary wavelets. If now an infinite plane wave front ABC is considered, as in *Figure 6.1*, travelling in the

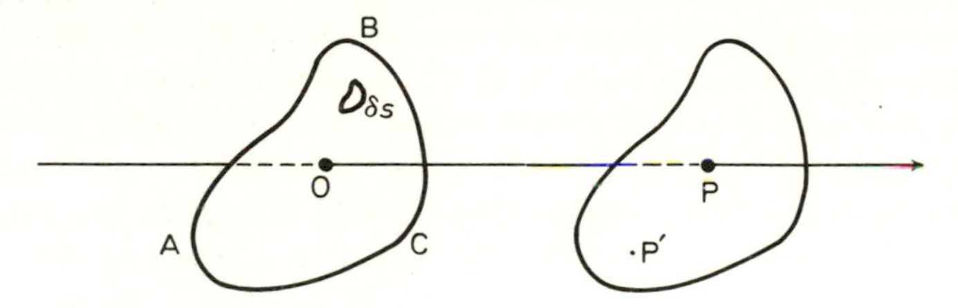

Figure 6.1. Propagation of a plane wave front

direction of its normal OP from a point source at infinity, then all points on this plane front will have the same phase, amplitude, and frequency. By Huygens' principle, a small area δs acts as a source of secondary wavelets and produces a disturbance at some point P. By the inverse square law and since the energy of the wave motion is proportional to the square of its amplitude, the amplitude of the disturbance produced at P is inversely proportional to the distance of P from the area δs and directly proportional to the area δs. To find the total amplitude of the disturbance at P owing to all such elemental areas δs, it is necessary to integrate over the total area of the plane wave front, taking into account that the areas are at different distances from P and therefore the wavelets arrive at P with different phases.

In the same way, the amplitude and phase can be found of the disturbance produced at some other point P′ lying on the plane through P parallel to the wave front. Since the plane wave front ABC is considered to be infinite in extent, the total disturbance produced at P′ or at any other point on this plane is identical to that produced at P. Hence a new plane wave front is produced through P, parallel to the original plane wave front ABC. Thus, by applying Huygens' principle to an infinite wave front, it is shown that light travels in straight lines. Similarly, a spherical wave is propagated as a spherical wave. A fuller treatment shows also that Huygens' secondary wavelets do not lead to the propagation of a back wave in a direction opposite to that of the wave front propagation and that the amplitude of the secondary wavelet is not constant in all forward directions. In a direction making an angle θ with the ray direction, the amplitude of the secondary wavelet varies as $(1 + \cos\theta)$. This is termed the *obliquity factor*.

Suppose now that a plane wave front is incident normally onto a screen which may be considered infinite in extent and that this screen has an aperture through which the wave may pass. If this aperture is very large compared to the wavelength of light so that it also may be considered infinite, the plane wave front will pass through to form a parallel-sided beam of light. If, however, the aperture is very small, a fraction of the wavelength of light, the part of the plane wave incident at the aperture is sufficiently small to behave as a single point source for Huygens' secondary wavelets. The wavelets then spread to form a spherical wave. For apertures between these two extremes, most of the light forms a plane wave front moving in a straight line – that is, as a parallel-sided beam of light – and the rest is propagated to the sides, into the geometrical shadow region. This apparent deviation of the light as it passes an obstacle, in this case the edge of the aperture, is termed *diffraction*. It is referred to as 'apparent' deviation because it is in fact not deviated but is the normal case of propagation from a Huygenian point source, as opposed to 'rectilinear' propagation, which is a special case occurring when the wave front may be considered as infinite in extent. The actual effect of an obstacle is not to deviate the light but to limit the area of the wave front, preventing the special case for rectilinear propagation.

An aperture, then, causes a redistribution of the light falling on a screen to produce a bright area approximately equal to that of the aperture. In the case of a circular aperture, this bright area is surrounded by a number of concentric rings, or fringes, rapidly decreasing in intensity with increasing radius. The larger is the aperture with respect to the wavelength of light, the nearer are the form and area of the bright area formed on a screen to those of the aperture, and the less is the energy removed from the bright area and distributed into the diffraction pattern.

Diffraction phenomena are the result of interference owing to wavelets arriving from different parts of a short wave front; interference phenomena are due to interference between two or more separate wave fronts. Optical diffraction phenomena may be divided into two classes: in *Fraunhofer diffraction* the incident light is effectively parallel and the diffraction pattern is observed on a screen which is effectively at an infinite distance; in *Fresnel diffraction* the wave fronts are spherical or cylindrical and so the source or screen or both are at finite distances. It is Fraunhofer diffraction that is of the more practical importance.

6.2 *Fraunhofer Diffraction due to an Aperture*

The simplest type of aperture to consider is a single slit. With this, Fraunhofer diffraction will occur with the arrangement shown in

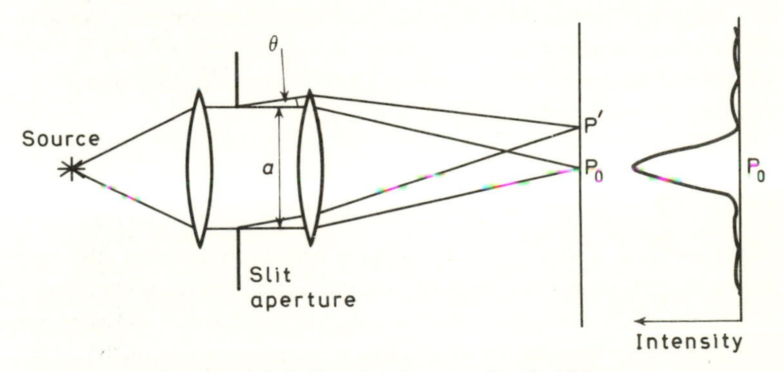

Figure 6.2. Diffraction due to a slit of width a

Figure 6.2 where the lenses are used to provide the required parallel rays and plane wave fronts. The intensity distribution produced on the screen is also shown in the figure. At the point P_0 on the screen, all the rays arriving are in phase. The sum of their amplitudes gives a maximum of intensity, the *principal maximum*, at this point, or more correctly along a line through P_0 parallel to the slit. Along some other line, say, through P′ on the screen, the rays from the elemental strip sources in the plane of the slit are inclined at some angle θ to the direction of the perpendicular to the plane of the slit. The rays arriving at P′ no longer have the same phase, since they have travelled different optical path distances, and hence their separate contributions to the resultant amplitude are not equal.

By summing the amplitude contributions from each of the elemental strip sources that may be imagined in the plane of the slit, taking into account their relative phases, it is found that the intensity distribution across the screen is given by

$$I = I_0 \frac{\sin^2 \beta}{\beta^2} \tag{6.1}$$

where I_0 is the intensity corresponding to the principal maximum at P_0 on the screen and β is a convenient variable used to describe the shape of the intensity distribution curve, where

$$\beta = \frac{\pi a \sin \theta}{\lambda} \tag{6.2}$$

The slit width a and direction θ are as indicated in *Figure 6.2*, and λ is the wavelength of the light. In particular, minima of intensity will occur on the screen whenever $\sin \beta$ becomes zero (equation 6.1), that is, whenever β becomes a multiple of π. The condition for a minimum of intensity is therefore (equation 6.2) when

$$\sin \theta = \frac{m\lambda}{a} \tag{6.3}$$

where $m = 1, 2, 3, \ldots$. For small θ, $\sin \theta$ becomes equal to θ. Between each of these minima, the intensity rises to form secondary maxima. The width of the principal maximum, that is, the distance between the first minimum occurring on either side, is twice the width of the secondary maxima. It should also be noted that the secondary maxima are not midway between the adjacent minima.

With a circular aperture, the diffraction pattern formed has rotational symmetry with a bright disc-like area centred on the axis, surrounded by light and dark rings. The central bright region or principal maximum is known as *Airy's disc* and contains approximately 84% of the light energy transmitted by the aperture. The first bright ring, that is, the second maximum, contains only 7% of the light. With a circular aperture, the successive maxima and minima are formed in a direction θ with the axis, where

$$\sin\theta = \frac{m'\lambda}{a} \tag{6.4}$$

where now m' is no longer an integer and the minima are no longer equally spaced. *Table 6.1* gives comparative values of m and m' for slit

Table 6.1 COMPARISON BETWEEN DIFFRACTION PATTERNS PRODUCED BY SLIT AND CIRCULAR APERTURES

	Slit aperture		*Circular aperture*	
	m	*Relative intensity*	m'	*Relative intensity*
Principal maximum	0	1	0	1
1st minimum	1	0	1·22	0
2nd maximum	1·43	0·0472	1·64	0·0175
2nd minimum	2	0	2·23	0
3rd maximum	2·46	0·0168	2·67	0·0042
3rd minimum	3	0	3·24	0

and circular apertures respectively. The relevance of diffraction at a circular aperture has already been discussed in Section 4.2 in relation to resolving power.

6.3 *The Diffraction Grating*

When the Fraunhofer diffraction pattern due to two parallel slits of equal width is observed by means of a lens and a screen, it is seen that the intensities in the pattern of parallel fringes appear to be modulated. Alone each slit would produce the same single-slit pattern as the other. When both slits are used the pattern does not remain the same but with double the intensity, but breaks up into narrower fringes caused by interference between the two diffraction beams. This pattern of narrow interference fringes has the general envelope shape of the single-slit pattern with the principal minima in the same positions. The intensity of each of the maxima, however, is four times that of the single-slit pattern at the corresponding point. *Figure 6.3* shows intensity distribution curves for single-slit and double-slit diffraction, where a is the slit width and b is the distance between the slits.

If now the number of slits is increased so that a grating is formed, the Fraunhofer diffraction pattern changes. The fringes become much sharper and so a grating with a dozen or so slits produces fringes that appear as lines. There is little difference in intensity between the fringes

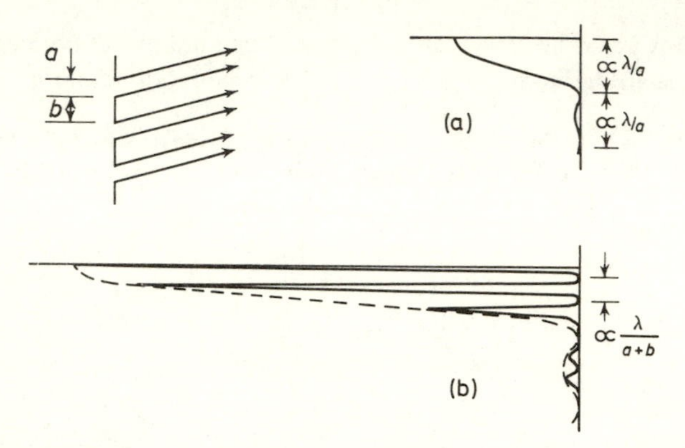

Figure 6.3. Intensity distributions in (a) single-slit and (b) double-slit diffraction patterns

in the central region so that the central maximum differs only very slightly from its neighbours. These fringes are all called the principal maxima. A number of secondary maxima are formed of much smaller intensity between each of the principal maxima, depending on the number of slits. All the fringes still, however, lie within a general envelope shape having the same width as the intensity distribution due to one slit *(Figure 6.4)*.

By summing the individual amplitude contributions in a direction θ from each slit, taking into account the phase difference between successive rays, it is found that the intensity distribution depends upon two factors: the first governs the intensity distribution pattern owing to diffraction from a single slit, and the second governs the interference pattern due to the multiple slits. The intensity variation with direction θ to the normal to the grating is

$$I = I_0 \frac{\sin^2 \beta}{\beta^2} \frac{\sin^2 (N\delta/2)}{\sin^2 (\delta/2)} \tag{6.5}$$

where β has already been defined for a single-slit diffraction pattern, i.e.

$$\beta = \frac{\pi a \sin \theta}{\lambda} \tag{6.6}$$

and δ is the phase difference between corresponding rays from successive slits and is given by

$$\delta = \frac{2\pi}{\lambda}(a+b)\sin\theta \tag{6.7}$$

where a is the slit width and b the distance between slits. N is the total

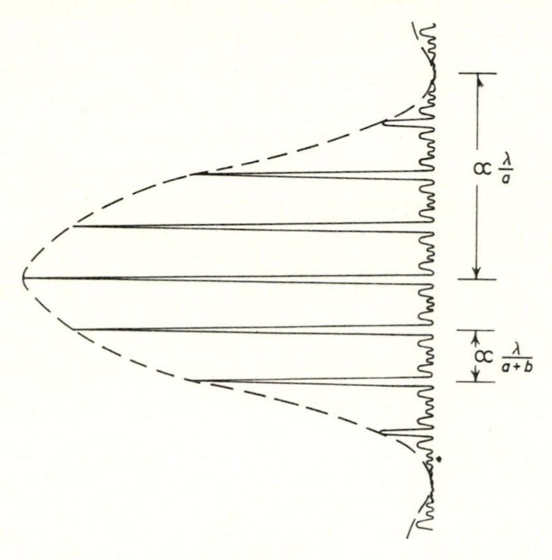

Figure 6.4. Intensity distribution across a six-slit diffraction pattern

number of slits in the grating. Of more practical importance, however, is the fact that a principal maximum occurs in a direction θ whenever

$$(a+b)\sin\theta = m\lambda \tag{6.8}$$

where $m = 0, 1, 2, \ldots$. Since also it is diffraction that gives the general envelope shape to the intensity distribution, it means that, when an interference maximum arising from interference between the successive slit contributions happens to coincide with a minimum due to diffraction, the diffraction factor is overriding and a *missing order* occurs. Minima occur due to the diffraction factor when

$$a\sin\theta = p\lambda \tag{6.9}$$

where $p = 1, 2, 3$. This is thus an overriding effect to equation 6.8.

6.4 *Formation of Spectra by a Grating*

Diffraction gratings are principally used as an alternative means of forming spectra. They have the advantage over prisms that absolute measurements of wavelength are possible without the need for calibration. When a slit aperture and a collimator are placed in front of a diffraction grating with the grating rulings parallel to the slit, principal maxima appear as sharp lines when viewed by a telescope. This is the same as replacing a prism on a spectrometer with a grating and using the spectrometer in the usual way. Equation 6.8 gives the directions θ of the principal maxima for light incident normally onto the grating. For the general case of light incident at an angle i with the normal to the grating, equation 6.8 may be written in the form

$$(a + b)(\sin i + \sin \theta) = m\lambda \qquad (6.10)$$

where $m = 0, 1, 2, \ldots$. This is the *grating equation*. With this, knowing the angle i which may conveniently be made zero, and the grating space $(a + b)$, the wavelength λ of light can be determined by measuring the angle θ with a spectrometer. The order, m, of the spectrum must be determined by inspection, starting with θ as a small angle.

The angular dispersion of a grating is the rate of change of the angle θ with change in wavelength. Obviously, the greater this dispersion, the greater is the separation between two spectral lines due to slightly different wavelengths – that is, the greater the resolution and accuracy in measurement of wavelength. To find the dispersion it is necessary to differentiate the grating equation (equation 6.10) with respect to λ, where angle i is a constant and independent of λ, so that the angular dispersion is given by

$$\frac{d\theta}{d\lambda} = \frac{m}{(a + b)\cos\theta} \qquad (6.11)$$

This means that the angular separation $d\theta$ between two spectral lines with a difference in wavelength of $d\lambda$ is directly proportional to the order m. Thus, although it may not be possible to resolve two lines in one order, by swinging the spectrometer telescope round it may be possible to find a higher order in which they can be resolved.

Since several orders are possible, the set of lines due to one order may overlap those due to another. Although this improves resolution because higher orders may be used, it may lead to confusion of orders and great care must be taken in deducing the correct order for a particular line.

As well as transmission gratings already discussed, *reflection gratings* work on exactly the same principle of introducing constant phase differences between successive rays. If the gratings are in step form, termed echelette gratings, most of the energy can be concentrated into one particular order. This avoids any danger of confusing orders.

Gratings are now manufactured by a replica technique proposed by T. Merton in 1948 and developed at the National Physical Laboratory, Teddington. A fine helix, which has been turned on a polished cylinder by a lathe, is used as the lead-screw of the lathe and combined with a long 'nut' made from pith clamped between metal shells. The soft 'nut' evens out periodic errors due to the original turning operation. With suitable gearing this lead-screw is used to cut a new helix of much finer pitch onto another polished cylinder, which in turn becomes the lead-screw for cutting an even finer pitched helix. This process may be repeated until a sufficiently fine, even-pitched helix is obtained. The required step form of the helix is determined by the form of the diamond cutting tool, which in this case does not remove a shaving as in conventional turning but displaces the metal into the correct form. Having obtained the cylinder, a thin plastic replica is made by coating the cylinder, allowing the plastic to cure and cutting longitudinally to remove. This in turn is replicated in gelatine which is then hardened and mounted on a glass plate. This transmission grating may be converted to a reflection grating by coating it with aluminium by vacuum evaporation.

6.5 Holography

With holography a three-dimensional image is recorded of a three-dimensional object. The technique, therefore, has interesting possibilities which have yet to be fully realised. If a lens is used to focus light reflected from an object onto a photographic plate, a two-dimensional replica of the object will be produced. Since it is a two-dimensional image, it appears the same when viewed from a slightly different angle. That is, all parallax effect is destroyed and with it all impression of depth or relative displacements. Without a lens the photographic plate would only record a diffraction pattern arising from light reflected from the object. This recorded diffraction pattern would vary in intensity but would contain no information as regards the relative phases of light reflected from the different parts of the object.

To produce a hologram, it is arranged that the photographic plate records both the diffracted light from the object together with light directly from the source, that is, light unmodulated by reflection from the object and thus able to act as a reference beam. The resultant image is due to interference between the two beams and will have a varying optical density which is now a function of both the intensity and phase of the light diffracted by the object. This image is a hologram. A suitable optical arrangement for producing such a hologram is depicted in *Figure 6.5*. Because the direct beam is required to act as a phase

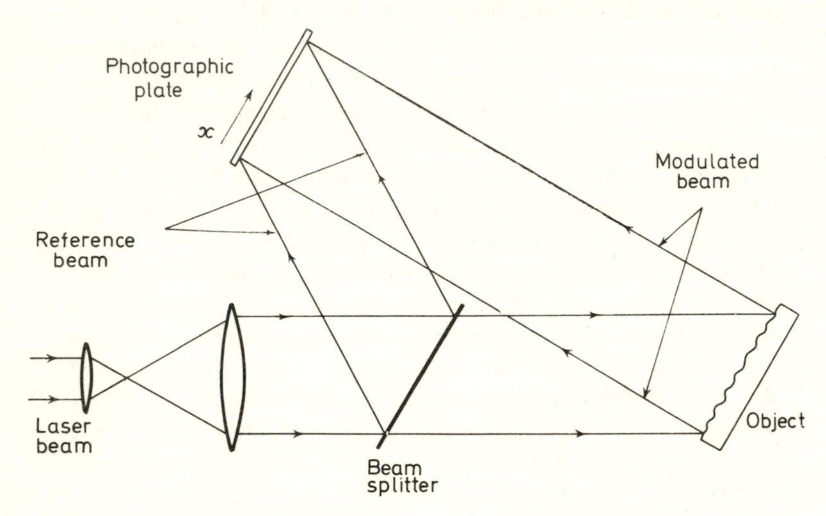

Figure 6.5: An optical arrangement for producing a holographic image of a three-dimensional object

reference beam for the diffracted beam, a laser source is required to provide the high degree of coherence and stability necessary. To view the image, the hologram must be illuminated by transmitted monochromatic light which generates further diffraction images which are in fact three-dimensional images of the original object. By changing the viewing angle a different aspect of the image is obtained. It is akin to viewing an object through a window – moving across the window reveals different aspects of the object but the range of viewpoints is limited by the window frame.

Referring again to *Figure 6.5*, it is seen that the reference beam is incident at a small angle onto the photographic plate. Over the area of the plate the amplitude A_0 of this beam is constant but its phase will vary linearly across the plate. That is, the phase at any point on the photographic plate can be written as αx, where α is a constant of proportionality. The complex amplitude of the reference beam at a point on the plate can, therefore, be written as $A_0 \exp(-\,\mathrm{i}\alpha x)$. The beam modulated by reflection from the object, however, will have an amplitude and phase that is a function of the shape and nature of the object. Taking into account variations in the object in the x-direction only for simplicity, the amplitude at a point on the photographic plate will be a function of x and may be written as $A(x)$. Similarly, the angle of incidence on the photographic plate will depend on the reflecting surface structure of the object. Hence the phase ϕ will also be a function of x. The complex amplitude of this beam, modulated by reflection from the object, may be written, therefore, as $A(x) \exp[-\,\mathrm{i}\phi(x)]$. Hence the total complex amplitude at a point on the photographic plate will be given by

$$A_{\mathrm{x}} = A_0 \exp(-\mathrm{i}\alpha x) + A(x) \exp[-\mathrm{i}\phi(x)] \qquad (6.12)$$

The intensity at a point x on the photographic plate is obtained by multiplying equation 6.12 by its complex conjugate, obtained by replacing i by $-$i. This is equivalent to squaring the amplitude. Hence

$$I_{\mathrm{x}} = A_0{}^2 + A(x)^2 + 2A_0A(x) \cos[\alpha x - \phi(x)] \qquad (6.13)$$

Thus the interference fringes formed on the photographic plate contain information that is a function of both the amplitude $A(x)$ and phase $\phi(x)$ arising from the modulation of the wave by the object. The plate, now termed a hologram, thus contains all the information relating to the shape of the three-dimensional object, in the form of a positive transparency.

To reconstruct a three-dimensional image of the object, it is necessary to illuminate the hologram with parallel coherent light. A wave front will then be formed which is of constant phase but has an amplitude that is a function of I_{x} in equation 6.13. The photographic emulsion will decide the form of this function which, assuming the exposure was correct so as to utilise the linear portion of the characteristic curve, will be dependent on the gamma of the film. If, for simplicity, a constant transmission factor, β, is assumed, the reconstructed amplitude, $A_{\mathrm{R}}(x)$, can be written as

$$A_{\mathrm{R}}(x) = \beta\{A_0{}^2 + A(x)^2 + 2A_0A(x) \cos[\alpha x - \phi(x)]\} \qquad (6.14)$$

The first two terms on the right-hand side of this equation describe a beam that is undeviated from the direction of the illuminating wave front but with some diffraction owing to the $A(x)$ term. The final

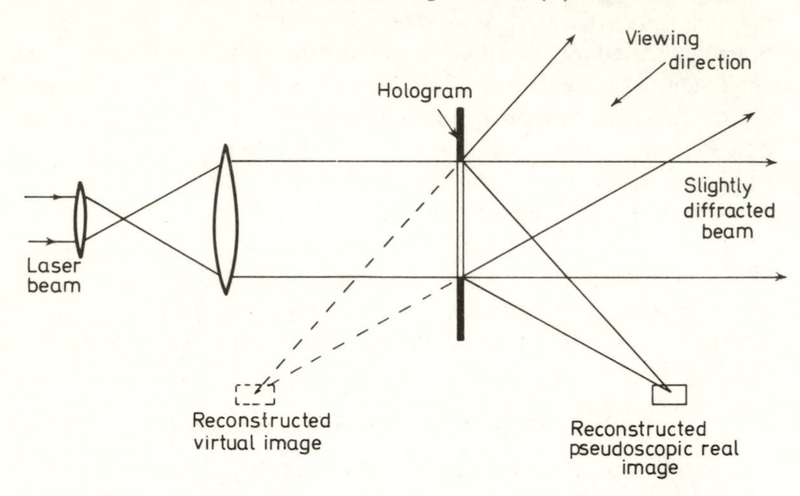

Figure 6.6. Reconstruction of three-dimensional images in holography

term, which includes information on both the phase and amplitude of the original beam as modulated by the object, can be expanded as

$$A_0A(x)\exp\{\mathrm{i}[\alpha x-\phi(x)]\} + A_0A(x)\exp\{-\mathrm{i}[\alpha x-\phi(x)]\} \tag{6.15}$$

Here the first term, which is a function of the original reference beam and the beam modulated by the object, describes a reconstructed beam that is deviated to one side to form a virtual image of the original object, as in *Figure 6.6*. Similarly, the second term describes a transmitted beam that is modulated by the original beam from the object and is deviated to the other side to form a real image of the object. However, this real image is generally of less use than the virtual image as its parallax relations are reversed to those of the original object, that is, it is a pseudoscopic image. Therefore it is generally the virtual image that is viewed, as in *Figure 6.6*.

When viewed in this way, changing the viewing direction gives a different viewpoint of the three-dimensional image. An interesting feature then arises if the hologram plate is broken into small pieces. Since, for a particular viewing direction, information relevant to the whole image comes only from a small part of the plate, that same view would still be seen if all the rest of the hologram were removed. That is, the reconstructed image can be seen through each of the fragments of the plate but only that aspect relevant to the particular fragment is seen. Again it is akin to viewing the object through a window but now the window has become very small and thus allows only one viewpoint.

X-RAY DIFFRACTION

6.6 Diffraction of X-rays

It has already been seen that light can be diffracted by a set of regularly spaced scattering objects, such as a diffraction grating, the only requirement being that the spacing of the diffracting objects should be the

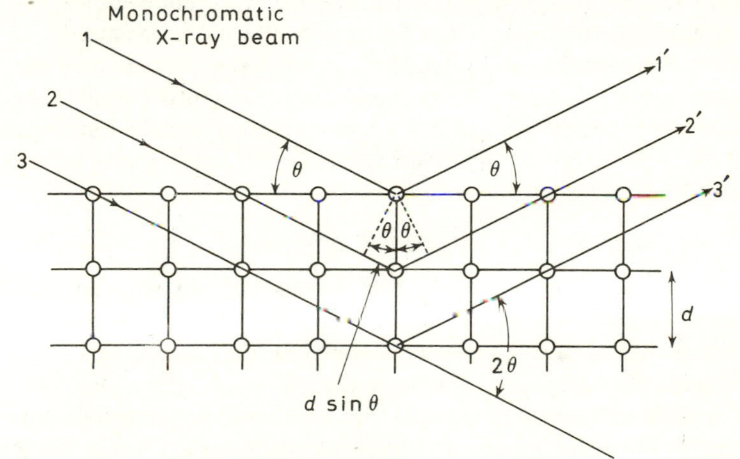

Figure 6.7. Diffraction of X-rays by a crystal

same order of magnitude as the wavelength of the diffracted wave. In a crystalline structure, the spacing between the layers of atoms is of the same order of magnitude as the wavelength of X-rays and it should therefore form a suitable diffracting object for X-rays. That diffraction is possible was first demonstrated by von Laue in 1912; following his work, W. L. Bragg derived the necessary conditions for X-ray diffraction.

In *Figure 6.7*, a parallel beam of monochromatic X-rays is incident at an angle θ onto a crystal. The angle θ, termed the Bragg angle, is measured with respect to a plane of atoms and not from the normal as is generally the case in optics. On striking the atoms, X-rays will be scattered in all directions but, in particular, they will be scattered in the particular direction shown in the figure. Between rays 1 and 2 there is introduced a path difference of $2d \sin \theta$, where d is the lattice spacing, the distance between parallel layers of atoms. If now this path difference is equal to a whole number of wavelengths, $n\lambda$, then rays $1'$ and $2'$ will still be in phase and will mutually reinforce each other. Similarly other rays, such as $3'$, will also be in phase with rays $1'$ and $2'$ and will add to the reinforcement. For all other directions of the scattered rays, the rays will be out of phase and destructive interference will occur. Thus the condition for the diffracted rays to be in phase and give reinforcement is given by Bragg's law

$$2d \sin \theta = n\lambda \tag{6.16}$$

where n is the order of reflection and is equal to the number of wavelengths in the path difference between rays reflected from adjacent planes. For a maximum of intensity in the direction θ, n can have any integral value consistent with $\sin \theta$ being less than unity. Owing to the values of the lattice spacings that occur in actual crystals and the repeat nature of the lattice, the order n may be taken as unity in practice and hence Bragg's law may be written as

$$2d \sin \theta = \lambda \tag{6.17}$$

It is also seen from *Figure 6.7* that the beam of maximum diffracted intensity is deflected through an angle 2θ relative to the direction of the incident beam. This angle 2θ is termed the diffraction angle and is the one usually measured experimentally.

At first sight it may be thought that the condition of equation 6.16 is due to an optical 'reflection' from layers of atoms but it is in fact due strictly to a diffraction phenomenon. That the processes are fundamentally different is seen from the fact that optical reflection can occur at any angle, whereas a Bragg reflection as given by equation 6.16 can only occur at specific angles. Also optical reflection is a surface phenomenon whereas the diffraction of an X-ray beam by a crystal is due to a reinforcement of all the rays scattered by the individual atoms within the crystal. Thus, although it is common usage to speak of 'reflecting' planes within the crystal, it is really diffraction that is implied. It should, however, be noted that, like optical reflection, the incident beam, the normal to the reflecting plane of atoms, and the diffracted beam are all in the same plane. True total reflection of X-rays can, however, occur with beams at glancing incidence but this is of little practical interest.

The importance of X-ray diffraction arises from the fact that a particular set of *d*-spacings is characteristic of a particular crystalline substance. By using X-rays of a known wavelength and measuring the angles 2θ of the various diffracted beams, values of the *d*-spacings can

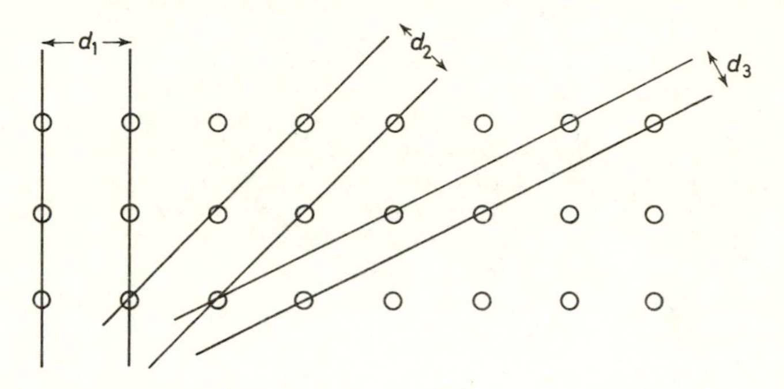

Figure 6.8. Illustrating the multiplicity of lattice spacings that exist within a crystal

be calculated to give structural information. That even a simple atomic arrangement can lead to a whole set of *d*-spacings is readily seen from *Figure 6.8*.

6.7 *The Crystal Lattice*

Owing to the repeat nature of the atomic arrangements within a crystalline structure, it is possible to consider a unit cell which may be imagined as the basic building block of the crystal. The crystal can be imagined to be divided up by three sets of planes, with the planes in each set being parallel and equally spaced. A set of parallelepiped cells will thus be formed such that each cell is identical with all of its neighbouring cells in size, shape, orientation, and arrangement and nature of contained atoms. The corners of these cells, formed by the intersection of the three sets of planes, constitute a point lattice, as in *Figure 6.9*. Any one of these identical cells may be designated a *unit cell.* Taking one corner of the unit cell as origin, the edges of the unit cell define the *crystallographic axes*. The *lattice constants*, in terms of the lengths a, b, and c and the angles between them α, β, and γ, then describe the size and shape of the unit cell.

The shape of the unit cell describes the particular crystallographic system. For example, if all the angles in the unit cell are 90° and the

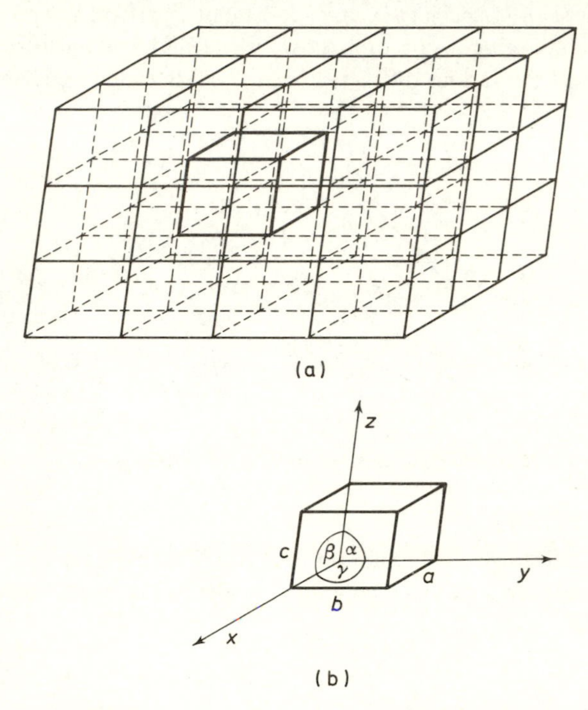

Figure 6.9. (a) A point lattice and (b) a unit cell

lengths a, b, and c are all equal, then the crystal lattice is said to be cubic. In all, only seven classes are needed to include all possible crystalline lattices. As well as a point lattice being composed of points at the corners of the seven forms of unit cells, a point lattice may be formed if it fulfils the basic requirement that each point has identical surroundings. With this basic requirement, there are in fact fourteen possible point lattices, termed *Bravais lattices*. Thus, in the cubic system, three Bravais lattice arrangements are possible, as is shown in *Figure 6.10*. The seven crystal systems and fourteen Bravais lattices are listed in *Table 6.2*.

To be able to refer to a particular set of reflecting planes it is

Table 6.2 CRYSTAL SYSTEMS AND BRAVAIS LATTICES

System	*Axial lengths and angles*	*Bravais lattice*
Cubic	$a = b = c, \alpha = \beta = \gamma = 90°$	simple body-centred face-centred
Tetragonal	$a = b \neq c, \alpha = \beta = \gamma = 90°$	simple body-centred
Orthorhombic	$a \neq b \neq c, \alpha = \beta = \gamma = 90°$	simple body-centred base-centred face-centred
Rhombohedral	$a = b = c, \alpha = \beta = \gamma \neq 90°$	simple
Hexagonal	$a = b \neq c, \alpha = \beta = 90°, \gamma = 120°$	simple
Monoclinic	$a \neq b \neq c, \alpha = \gamma = 90° \neq \beta$	simple base-centred
Triclinic	$\mathrm{a} \neq \mathrm{b} \neq \mathrm{c}, \alpha \neq \beta \neq \gamma \neq 90°$	simple

necessary to be able to label them in some way and this is achieved by the Miller notation. To represent a crystallographic plane or one parallel to it, the plane is imagined extended to cut the x, y, and z axes defining the edges of the unit cell, as shown in *Figure 6.11*. The intercepts of the plane with these axes are measured in terms of units of a, b, and c, the unit cell dimensions. Reciprocals of these intercepts are then taken and the fractions cleared to give the smallest integers. For example, if in *Figure 6.11*, a, b, and c define the dimensions of the unit cell, measured along the axes x, y, and z as shown, then the plane ABC may be described by the Miller notation as follows. The intercepts OA, OB, and OC on the x, y, and z axes respectively are 1, 2, and 2 measured in units of a, b, and c respectively. The reciprocals of the intercepts are 1, ½, and ½ and multiplying through to remove the fractions gives the Miller

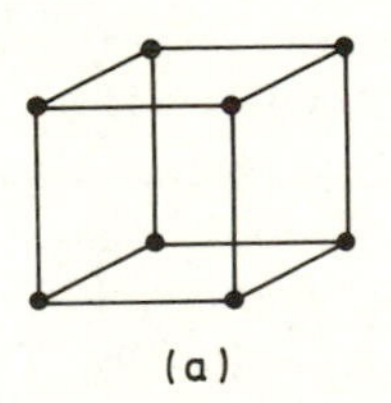

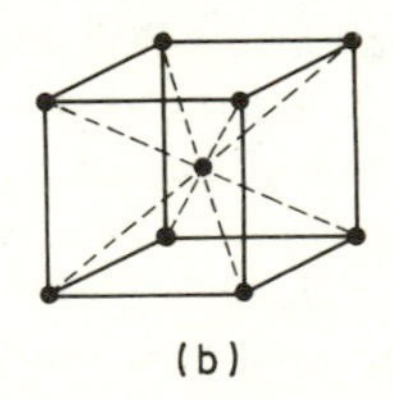

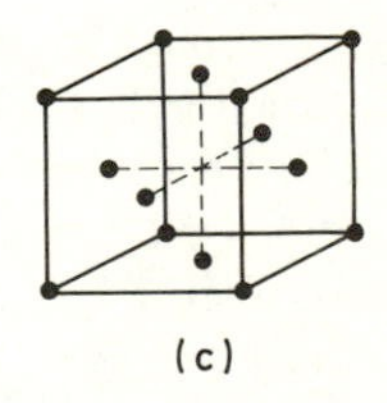

Figure 6.10. (a) Simple cubic, (b) body-centred cubic, and (c) face-centred cubic Bravais lattices

indices (211) of the plane ABC. Because of the repetitive nature of the structure, any plane parallel to this will be described by the same indices. Negative intercepts are indicated by a bar over the appropriate index. A

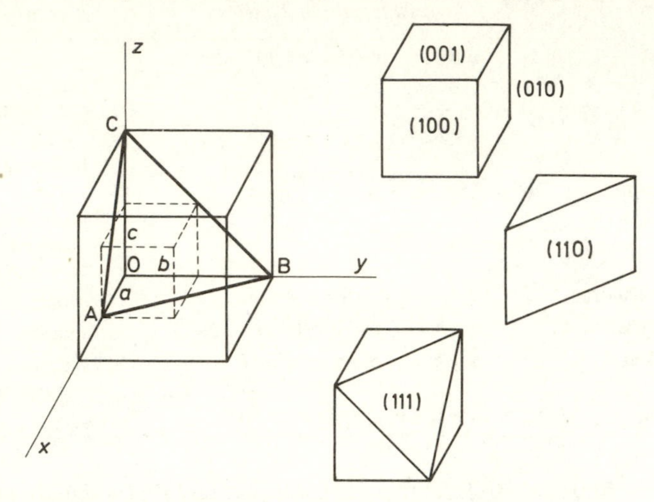

Figure 6.11. Miller indices for various crystallographic planes

plane parallel to, say, the *xz*-plane and intercepting the *y*-axis at two units would have intercepts on the three axes ∞, 2, and ∞. Taking reciprocals gives 0, ½, and 0 and multiplying through to obtain the lowest set of integers gives the Miller indices (010) for the plane.

A direction within the crystal may be indicated by a similar method of notation. The direction is first described in terms of its vector components resolved along each of the coordinate axes *x*, *y*, and *z*. These components are again measured in terms of the dimensions *a*, *b*, and *c* of the unit cell. For example, the direction of the line OA in *Figure 6.12* is described by the vector components $\frac{3}{2}$, 2, and 1 measured along the *x*, *y*, and *z* axes and in units of *a*, *b*, and *c* respectively. Multiplying through to obtain the smallest integral coefficients gives the Miller direction indices [342]. Again a negative direction may be indicated by a bar over the appropriate index. Also, as before, owing to the repetitive structure of the crystal, this is the direction of any other line parallel to OA. The use of the different shaped brackets should be noted – (*hkl*) indicates a plane and [*uvw*] indicates a direction. The notation {*hkl*} may be used to indicate the whole set of planes (*hkl*), (*khl*), ($\bar{h}kl$), etc. These are not parallel planes but they belong to the same family. In a cubic crystal all the planes in the same family have identical arrangements of atoms and consequently the same character. Similarly, the notation $\langle uvw \rangle$ is used to indicate the set of directions

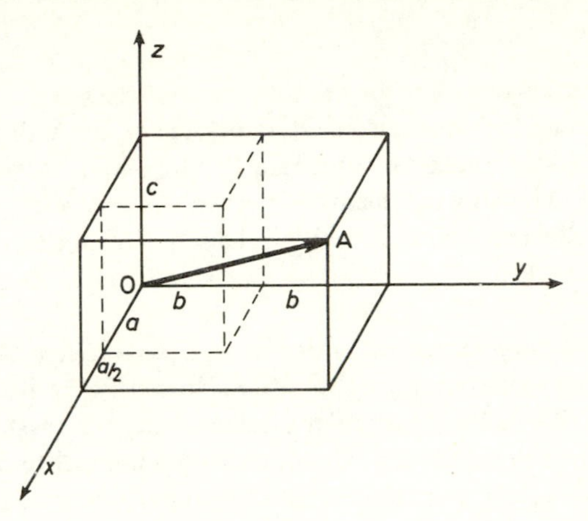

Figure 6.12. Miller indices for directions within a crystal lattice

$[uwv]$, $[wuv]$, etc. For a cubic system it will be seen that the direction $[hkl]$ is perpendicular to the plane (hkl).

6.8 X-ray Diffraction Methods

By suitably setting up a crystal in an X-ray beam, the directions of the diffracted beams can be determined experimentally and the respective d-spacings determined from the Bragg equation, equation 6.17. From this information it is possible to determine the crystallographic class and the size of the unit cell. For example, if it is found that the set of d-values each satisfies the equation

$$\frac{1}{d^2} = \frac{h^2 + k^2 + l^2}{a^2} \tag{6.18}$$

where h, k, and l are integers, then the crystal is cubic with a unit cell size a. Equation 6.18 thus predicts all the possible Bragg reflections that can occur from the planes (hkl). For a tetragonal crystal the equivalent expression is

$$\frac{1}{d^2} = \frac{h^2 + k^2}{a^2} + \frac{l^2}{c^2}$$

and for a hexagonal crystal

$$\frac{1}{d^2} = \frac{4}{3}\left(\frac{h^2 + hk + k^2}{a^2}\right) + \frac{l^2}{c^2}$$

Similarly, expressions exist for the other crystal classes.

To determine the positions of the individual atoms within the unit cell, it is necessary to take into account the intensities as well as the directions of the diffracted beams. However, the specialist techniques of structure determination are beyond the scope of this text and the reader is referred to any of the many specialist texts on X-ray crystallography.

Apart from purely structural information, X-ray diffraction techniques can yield other useful information. To satisfy the Bragg condition, $2d \sin \theta = \lambda$, for a particular d-value, either the crystal must be orientated in every conceivable direction with a monochromatic X-ray beam or, with a fixed orientation, the crystal must be irradiated with a range of wavelengths, that is, 'white' radiation.

In the *Laue method*, a single crystal is fixed in one position and irradiated by white radiation. There are two variations of the method, depending on the position of the photographic film relative to the crystal and the incident beam. The two methods are illustrated schematically in *Figure 6.13* where it is seen that the film is always in the form of a flat plate placed normally to the incident beam. In both positions the film has a central hole to allow the direct beam to pass through to prevent excessive blackening of the film. A suitable fine beam of X-rays is defined by a two-hole collimator system to give an X-ray beam of about 0·5 mm in diameter. The X-ray tube is operated to give a high-intensity continuous radiation. Whenever the orientation and wavelength conditions satisfy the Bragg equation for a particular d-value, a Bragg reflection will occur, producing a black spot where the particular diffracted ray intersects the film. The spots will lie on curves which are generally ellipses or hyperbolas for transmission patterns, or hyperbolas for back-reflection patterns. From an examination of the film and the geometry of the arrangement, the axes of the cones of the diffracted rays can be deduced and hence the orientation of the crystal lattice found. This then is the main use for Laue methods: for determining the orientation of a crystal or, as is perhaps more usual, to enable the crystal to be orientated into a particular direction. The shape of the spots also gives information concerning the perfection of the lattice. If the lattice is strained, then the spots will be smeared out or their shape distorted.

The other main use of X-ray diffraction as a tool, other than its use for orientation, is with the *powder method*. If the crystal is ground up into a fine powder and irradiated by a monochromatic X-ray beam, every particle of powder acts as an individual crystallite. Owing to the large number of crystallites some of these will be suitably orientated to produce a Bragg reflection. In fact every set of lattice planes will be capable of producing a reflection. In the case of a metal, the grains will generally be of random orientation and a piece of wire will often already be in a suitable condition for powder diffraction examination. In the *Debye–Scherrer method* (*Figure 6.14*) the ground-up specimen

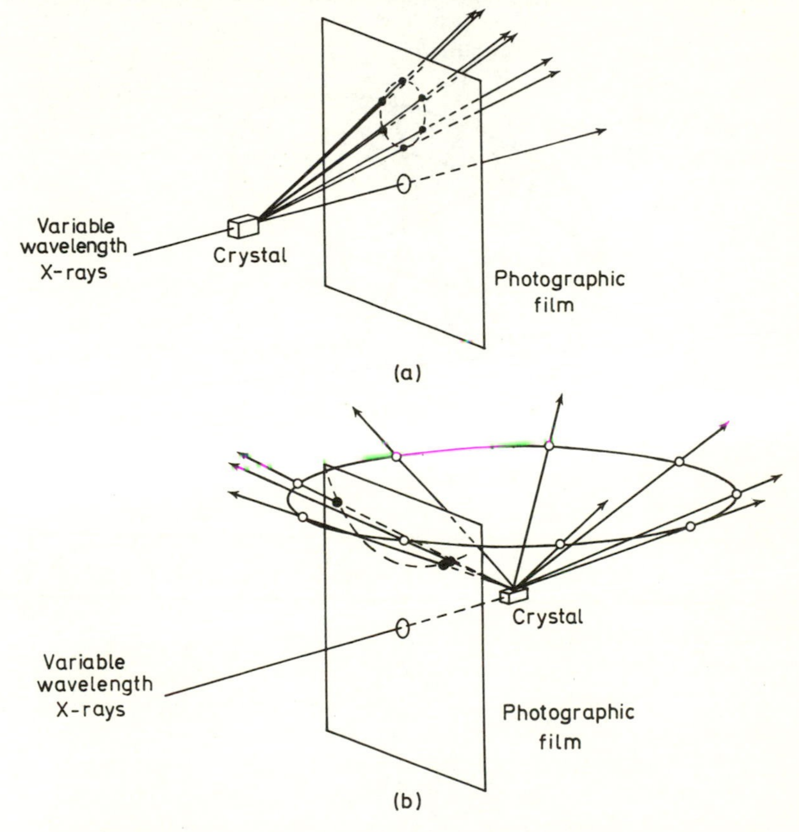

Figure 6.13. Formation of X-ray diffraction patterns by (*a*) *transmission and* (*b*) *back-reflection Laue methods*

is mounted at the centre of a cylindrical strip film camera; camera diameters vary from about 50 mm to 200 mm. The powdered specimen can be coated onto a hair coated with gum or mixed with gum tragacanth and rolled between glass plates into a cylinder. A specimen of about 0·5 mm in diameter and length is adequate, although a longer specimen makes the mounting easier. However, it is possible to deal with much smaller specimens. The specimen must be accurately centred on the camera axis. To increase the probability of satisfying the Bragg condition for all the possible reflecting planes, the cylindrical powder specimen is rotated about its axis at about two revolutions per minute during the period of the exposure, which may perhaps be half an hour but will be very dependent on the nature of the specimen. A particular plane that satisfies the Bragg condition will diffract a ray in a direction

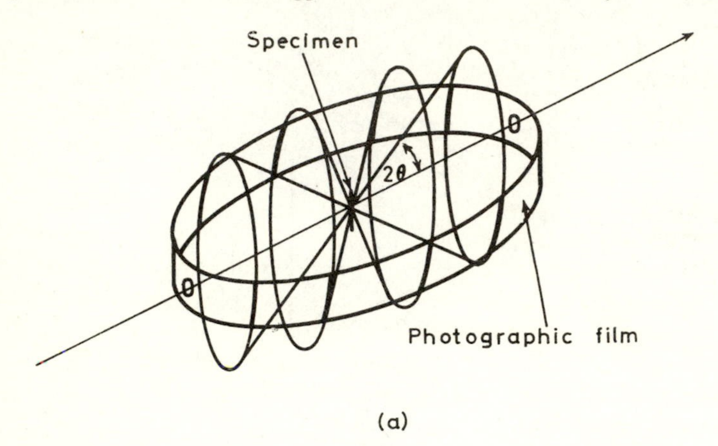

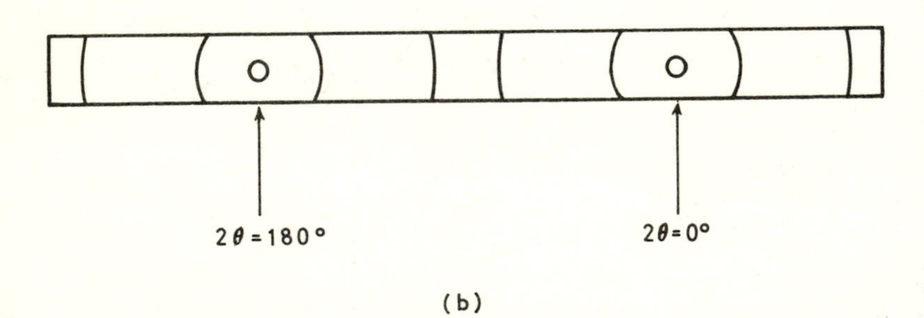

Figure 6.14. (a) Debye–Scherrer powder method, showing (b) cylindrical film laid out

2θ from the main beam. Owing to the random orientations of the crystallites within the specimen, the diffracted rays will form a cone of semi-angle 2θ. The intersection of this cone with the film gives two lines symmetrically placed with reference to the main beam. The lines will be curved except for lines that occur exactly at $2\theta = 90°$, when they will be straight. In fact the pattern produced on the film is a series of spots – one from each crystallite – sufficiently close together to appear as continuous lines. After exposure and processing, the cylindrical film is laid flat and, from the dimensions of the camera and measurements along the length of the film, the value of 2θ corresponding to each Bragg reflection is determined and, knowing the X-ray wavelength, the *d*-values are calculated from equation 6.17. The characteristic wavelength used should not, however, be shorter than the K absorption edge of the specimen. A camera diameter of 57·3 mm is often used, as a distance along the film of 1 mm is equivalent to an angle $\theta = 1°$.

An important use of this method is in the identification of compounds. For example, chemical analysis may determine the presence of iron in a sample and an iron oxide may be deduced. However, the precise form of the oxide cannot be determined, especially if the sample is of mixed oxides. Since each oxide has a different crystalline form and will therefore produce a different X-ray diffraction pattern, the X-ray powder method is able to differentiate between, for example, FeO, Fe_2O_3, and Fe_3O_4. The method also has the added advantage that the specimen is not altered or destroyed and is therefore available for further study. The set of *d*-values is, in fact, virtually unique to a particular chemical compound and it is this that makes the method such a powerful tool for analysis.

To facilitate the identification of an unknown compound, the American Society for Testing and Materials, ASTM, has issued data for a very large number of compounds. To identify an unknown compound by means of the ASTM Index, all the *d*-values are determined for the powdered specimen and an estimate is made of the intensity of blackening of each of the lines produced on the film. The strongest line is allocated an intensity of 100 and the other lines are allocated intensities in proportion to this. A visual estimate is generally sufficiently accurate and with experience can be made in 5% steps. The ASTM Index is based on the three strongest lines and is used by locating the three strongest lines of the unknown pattern in the Index. Often several groups of three strongest lines in the Index agree fairly closely in both *d*-values and intensities with those of the unknown pattern. The Index then quotes references to the main body of the ASTM Data File for each group of three strongest lines. Here all the lines of a pattern are given, with both *d*-values and intensities, and from one of these references a unique match to all the lines of the unknown pattern can generally be made and the sample thus identified.

The system becomes more complex to use if the unknown sample is a mixture of several compounds. Then each compound produces its own diffraction pattern and there is the possibility of, say, two weak lines coinciding to give a strong line. This may confuse the choice of the three strongest lines for the Index search and it may become necessary to search for all permutations of three lines and check each reference with the main Data File. However, with experience and some prior knowledge of the sample, so that compounds known not to be present can be eliminated, samples containing mixtures of a number of different compounds can be quickly and accurately analysed. By making up a range of known mixtures and obtaining their powder diffraction patterns, estimates of percentage composition of the sample under investigation can be made by comparison. The method, however, is not suitable for less than about a 5% impurity in a two-component mixture.

As well as its use for identification of compounds, a powder pattern can give information as to grain size which is of especial interest in metallurgy. A small metal specimen can often be regarded as a powder specimen since it is made up of grains that are generally of random orientations. If the grains are large, the patterns produced on the film will not be of continuous lines since there will be insufficient crystallites in the X-ray beam to satisfy all possible Bragg reflections for every orientation about the direct beam. Thus, for comparatively large grain sizes, the patterns are spotty; below a grain size diameter of about 10^{-5} to 10^{-6} m, the powder diffraction patterns lose their spottiness and become continuous lines. For grain sizes below about 10^{-7} m, the lines become broadened. A strained specimen may also give line broadening and it may then be necessary to anneal the specimen before study. This is especially so of metallurgical specimens that may have been produced by filing or by the turning of a small rod from a larger sample. Metallurgical specimens may also show preferred orientations. For example, in a drawn wire, most of the grains may have a particular crystallographic direction parallel to the wire axis or, in a rolled sheet, parallel to the rolling direction. Then the condition for all possible orientations is not satisfied and the X-ray diffraction patterns will not be continuous lines.

As well as photographic recording of the diffraction patterns, recording by counters is also used. In an automatic X-ray diffractometer, shown schematically in *Figure 6.15*, the specimen is irradiated by a parallel beam of monochromatic X-rays. The specimen in the form of a slab is rotated about the diffractometer axis. Any Bragg reflections that occur as the specimen is rotated will be recorded by a counter, for example, a Geiger counter, which is traversed about the same axis but at twice the angular speed of the specimen. Thus the counter is always

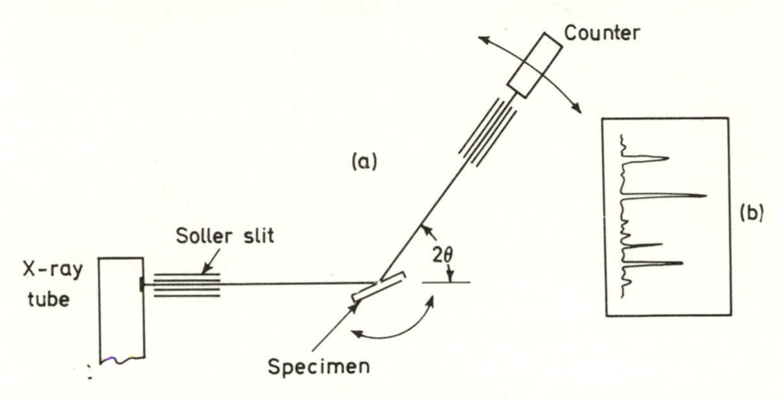

Figure 6.15. Arrangement of an X-ray diffractometer (*a*) *with output from counter being traced by a chart recorder* (*b*)

in the correct position should a Bragg reflection occur. The output from the counter is handled electronically and displayed on a chart recorder so that both 2θ angles and intensities can be read directly from the recorded trace. To produce the parallel beam of X-rays, a Soller slit is used, which is just a set of thin metal plates separated by about 0·5 mm which reduces the unwanted divergence of the X-ray beam to about 1·5°. The specimen should be finely ground as before but, instead of forming it into a fine rod, it must now be formed into a sheet. It can be smeared onto the flat specimen holder, if necessary incorporating some gum as a binder, or can be packed into a rectangular depression in a specimen holder and the surface scraped flat. However, more sample is required than in the Debye–Scherrer powder camera method – sufficient to cover an area of perhaps 8 mm x 15 mm, which may cause difficulties. Metal specimens can, of course, be mounted in sheet form.

To make a complete traverse of the 2θ angles, allowing time to accumulate sufficient counts to give an accurate recording of each of the diffraction lines, takes about the same time as is required to record the same pattern on film using a powder camera. For routine work in which it may only be necessary to establish the presence or intensity of one particular diffraction line to categorise the sample, then a diffractometer has an obvious advantage in time. The accuracy achieved using a diffractometer, both for angle measurement and consequently d-values and for intensities, is, however, considerably greater than by photographic techniques. Against this of course must be balanced the considerably greater cost of a diffractometer, which is due to the accurate gearing required to maintain the correct orientations of specimen plane to counter and the stabilised electronics required to achieve accurate intensity measurements.

ELECTRON DIFFRACTION

6.9 Electron Diffraction Methods

Mention has already been made in Section 4.8 of the use of an electron microscope for electron diffraction studies. The principles are the same as those used in X-ray diffraction and the same type of information is determined. The electron diffraction pattern obtained using an electron

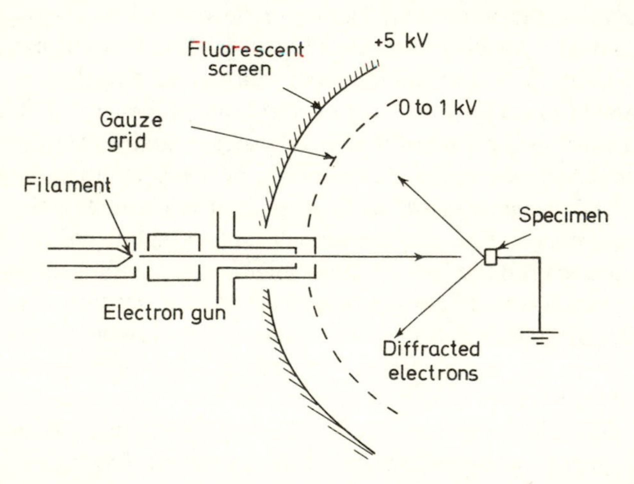

Figure 6.16. Arrangement of electrodes to produce low-energy electron diffraction

microscope is equivalent to that produced with X-rays by a transmission Laue method. The nature of an electron microscope specimen means that a concentric ring pattern is most generally obtained, although single-crystal spotty patterns do occur. Identification of unknown compounds is not quite so simple as with X-rays as the electron wavelength and the magnification are less precisely known. It may therefore be necessary to use a known sample to obtain a suitable calibration factor. Otherwise information as to crystal orientation and grain size may be obtained as with X-ray methods.

In obtaining an X-ray diffraction pattern, use is made of the penetrating properties of the X-rays themselves. If, however, it is required to study surface layers only, this penetrating property is a disadvantage. With electrons the penetration can be controlled by slowing down the electrons by suitable electric fields. In *Figure 6.16* is shown an arrangement of electrodes to produce low-energy electron diffraction (LEED). A low-energy beam of electrons impinges on the specimen to produce electron diffraction by back reflection. Owing to the low energy of the electron beam, there is very little penetration so that the diffraction pattern results only from the outermost layers of atoms. The back-diffracted electron beams are then accelerated by gauze grids to pass through and impinge on a fluorescent screen, when the resulting bright spot pattern may be photographed. It is necessary to have a high positive potential on the screen to attract the electrons with sufficient energy to excite fluorescence. As well as giving structural information from the positions of the spots in the pattern, variations in intensity can yield data on the amplitude of vibration of surface atoms. However, the analysis of the diffraction pattern does present problems and is somewhat uncertain.

An alternative technique for the examination of surface structure is to use reflected high-energy electron diffraction (RHEED). With this method, a high-energy electron beam is incident onto the specimen at a glancing angle. The beam is then diffracted only by those atom layers near to the surface to produce a diffraction pattern consisting of a series of lines on a fluorescent screen. The techniques of LEED and RHEED can be regarded as complementary.

Seven

SPECTROSCOPY

7.1 Optical Spectra

If white light is passed through a prism or a diffraction grating, it will be dispersed into its constituent colours. The spectrum formed is a *continuous spectrum* since it is a continuous band of colour spreading from violet through to red, covering the wavelength range of approximately 4×10^{-7} to 8×10^{-7} m. Other units of wavelength are the angstrom, Å, equal to 10^{-10} m, and the millimicron, mμ, and micron, μ, equal to 10^{-9} m and 10^{-6} m respectively. Also commonly used is the wave number, the reciprocal of the wavelength measured in centimetres. The spectrum may also spread into the non-visible region – into the infrared, where the wavelengths are greater than the red of the visible region, and into the ultraviolet, where the wavelengths are shorter than the blue end. The ultraviolet region of the spectrum may be subdivided into two spectral regions – between 2×10^{-7} m and 4×10^{-7} m, termed the near-ultraviolet region, and below 2×10^{-7} m, termed the far or vacuum ultraviolet.

This type of continuous spectrum is emitted by all incandescent solids and some dense gases. As a solid is heated, it first emits heat in the form of infrared radiation. As its temperature rises, it becomes 'red-hot' when it emits wavelengths extending into the red end of the visible spectrum. At higher temperatures, it becomes 'white-hot' when it then emits wavelengths extending into the blue end of the visible spectrum which, with the mixture of other wavelengths present, make up white light. Such emission of the various wavelengths is due to the thermal vibration of the molecules within the solid. The wavelength of light emitted is related to the frequency of vibration of the molecules in the hot solid but, owing to collisions with neighbouring molecules, there can be no specific frequency of vibration and so white light is emitted.

If the spectrum caused by burning organic compounds or, generally, the spectrum obtained by suitably excited freely moving molecules as opposed to atoms is examined, a *band spectrum* is seen. This appears as broad bands which are more intense at one edge and fade away towards the other edge. On closer examination it is found that these bands have a very fine line structure.

Whereas band spectra are due to freely vibrating molecules – that is, as gases or vapours – *line spectra* are due to vibrations of free atoms. The line spectrum of an element is a characteristic of that particular element and is normally only produced by an element in its gaseous state. A line spectrum is formed when the atomic structure of an atom has been disturbed and is due to the radiant energy emitted as the electrons return to their stable energy levels, that is, it is due to electronic changes. If two carbon electrodes are brought together and separated to form an arc, the glowing electrodes themselves emit a continuous spectrum, whereas the arc which contains vaporised electrode material as well as the atmospheric gases emits a mixture of line and band spectra. An arc may be struck in a vacuum to exclude the atmospheric gases. If one of the electrodes is a metal or contains some metal, the arc contains its vapour and the line spectrum of the metal is emitted. Similarly, if an electrode is sparked onto a metal surface, the vapour emits the spectrum of the metal, although in some cases this spectrum may be different from that due to the arc. Thus it is usual to refer to both *arc* and *spark* spectra as appropriate.

The spectra referred to above are all examples of *emission* spectra. If white light – that is, a continuous spectrum as emitted by an incandescent solid – is passed first through a vapour and then into a spectrometer, it is found that the continuous spectrum is crossed by a series of dark lines or bands. These lines form the *absorption spectrum* of the vapour and are in the same positions as the lines or bands that are emitted by the vapour if heated. The cold vapour absorbs those wavelengths that it emits if hot. Like emission spectra, absorption spectra may be in the form of lines or bands. Line absorption spectra are obtained only with monatomic gases and vapours; band absorption spectra are formed by molecules. Absorption spectra may also be obtained by passing light through solids and liquids.

Apart from simple identification and the production of wavelength standards, emission spectra have little practical use in comparison with the information that can be obtained from absorption spectra. If a molecule is exposed to radiation of wavelength λ, the molecule will absorb radiation and its energy will increase. This increase in energy is given by the relation

$$E = h\nu = \frac{hc}{\lambda} \tag{7.1}$$

where ν is the frequency of the incident radiation, c the velocity of light, and h Planck's constant. The magnitude of this absorbed energy will decide the nature of the change produced in the molecule and may thus involve electronic, vibrational, or rotational changes. Relatively large quanta of energy are needed to produce electronic changes, whereas only small energies are required to produce rotational changes. Normally a molecule will be in its lowest energy, or ground, state and the absorption of energy will cause the molecule to be raised into one of its possible excited states. Thus the normal ground state electron shell structure of the molecule may be disturbed and the outer electrons moved into higher orbits to give excited electron states. In addition, each of these electron energy states has ground and excited vibrational states and each of these in turn has ground and excited rotational energy states. This means that, if a molecule absorbs energy from a source in the far-infrared region, only sufficient energy will be absorbed for its rotational energy to be affected. If the source is in the near-infrared region, transitions will be produced between rotational and vibrational energy levels of the lowest, that is, ground, electronic energy state. For radiations of greater energy, such as from ultraviolet sources, electronic, vibrational, and rotational changes can take place so that transitions between vibrational and rotational energy levels of different electronic states are possible. The ultraviolet absorption spectrum of a molecule will thus be more complex than its corresponding infrared spectrum. For polyatomic molecules the large number of closely spaced energy sublevels causes the ultraviolet absorption spectrum to appear as broad bands or band envelopes. It is only in some simple molecules, using refined techniques, that the fine structure of these bands can be measured and analysed. If an electronic transition is produced, the molecule may return to its ground state by giving up its excess energy in the form of heat or as a fluorescent radiation of a longer wavelength.

7.2 Applications of Optical Spectroscopy

Spectroscopy is much used for the identification of substances and the specification of their purity. This is done simply by comparison with known specimens and mixtures. However, it is absorption rather than emission spectroscopy that has found the wider range of applications. For a substance to be suitable for analysis by absorption spectroscopy, it should be a substance whose absorption is in agreement with the Beer–Lambert law. This states that the fraction of incident light absorbed by a substance is proportional to the number of molecules in the light path, so that if a substance is dissolved in a solvent the absorption by the solution will be proportional to its molecular concentration. This, of course, assumes that the solvent itself does not have any absorption in the particular wavelength region. The law can be stated as

$$\text{absorbance} = \log_{10}\frac{I_0}{I} = \epsilon c l \qquad (7.2)$$

where I_0/I is the ratio of the incident to transmitted light intensities, ϵ is the absorption coefficient, c the concentration, and l the length of the absorbing path. The absorbance is also known as the 'extinction' or the 'optical density'. According to the Beer–Lambert law, the absorbance should remain constant if the product of the concentration and the path length remains constant. In practice, this is not always so, owing for example to molecular association of the solute at high concentrations, or fluorescence effects. It is, therefore, necessary to test the validity of the law over the range of concentrations likely to be used.

To analyse a mixture of substances, application may be made of the Beer–Lambert law for each of the components independently. Providing the components absorb independently, the measured absorbance is given by

$$\begin{aligned}\text{absorbance} &= \log\frac{I}{I_0}\\ &= (\epsilon_1 c_1 + \epsilon_2 c_2 + \ldots)\, l \qquad (7.3)\end{aligned}$$

where $\epsilon_1, \epsilon_2, \ldots$ are the extinction coefficients of the various components at a given wavelength, and $c_1, c_2, \ldots$ are the corresponding concentrations. If there are n components in the mixture, absorbance measurements must be made at n different wavelengths and the resulting n simultaneous equations solved to obtain the concentrations.

The infrared absorption spectrum is important in the determination of molecular structure as it allows a qualitative analytical approach to be made. The method enables information to be obtained regarding the nature of the bonding between the atoms in the molecule. For example, a diatomic molecule A—B can only vibrate longitudinally along the bond. The vibrational frequency of the stretching of the bond A—B is given by

$$\nu = \frac{1}{2\pi c}\left(\frac{f}{m}\right)^{1/2} \tag{7.4}$$

where c is the velocity of light, f the force constant of the bond, and m the reduced mass of the system, given by

$$m = \frac{m_A \times m_B}{m_A + m_B} \tag{7.5}$$

where m_A and m_B are the individual masses of A and B. Equation 7.4 is analogous to Hooke's law applied to the oscillations of two bodies connected by a spring. Similarly, the vibrational frequencies of other degrees of freedom in more complex molecules may be studied. However, it should be borne in mind that, besides those factors of equation 7.4, there are a number of other minor factors that can affect the vibrational frequencies of bonded atoms.

By repeatedly scanning a wavelength range, spectrophotometers may be used to determine the beginning of a chemical reaction and the appearance of new products and so give information in the study of reaction kinetics. Other uses of absorption spectroscopy are in the determination of molecular weight and the study of *cis–trans* isomerism. Here the absorption spectra of the isomers will be different since they will differ in their spatial arrangement of groups about a plane. However, for details of these and other specialised chemical techniques, reference should be made to a text on chemical spectroscopy.

7.3 Spectrometers and Spectrophotometers

An arrangement of a simple spectrometer for the study of emission spectra is shown in *Figure 7.1*. Light from the source to be analysed, which could be a gas discharge lamp or an arc or spark, is focused onto the slit of a collimator. The collimator, a lens–slit combination, produces a parallel beam of light which is incident onto the prism face. The prism itself is mounted on a rotatable table which may be tilted so that the refracting edge of the prism is vertical and parallel to the source slit. In the visible region of the spectrum a glass prism would be used, whereas to obtain adequate transmission in the far-ultraviolet region calcium fluoride or lithium fluoride prisms would be used. Similarly, quartz prisms would be used for the near ultraviolet and rock salt (sodium chloride) for the infrared regions. Several other materials are also used, especially for different regions of the infrared spectrum. As well as the prism, the other optics of the system must be of suitable transmitting materials. Alternatively, a grating may be used in place of a prism, as discussed in Section 6.4.

The beam of light from the collimator is then refracted by the prism, or diffracted if a grating is used, as a series of parallel beams each at a

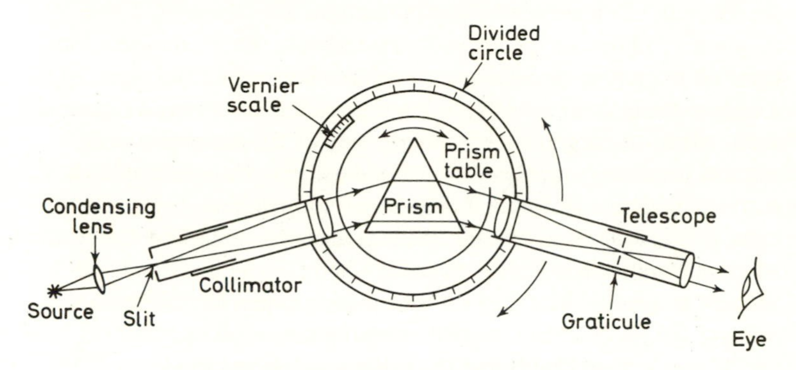

Figure 7.1. Arrangement of a spectrometer

slightly different inclination, depending on the wavelength. These rays are focused by a telescope so that each individual wavelength is brought to a separate focus. The telescope can be moved about a vertical axis passing through the centre of the table so that the angle between the axis of the telescope and the axis of the collimator may be read from the divided circle. By centring a particular spectrum colour (wavelength) on the centre line of the graticule in the telescope eyepiece, the deviation for that wavelength may be determined. By adjusting both the telescope and prism position, the position for minimum deviation may be found for each wavelength. To obtain a photographic record, the eyepiece is removed and replaced by a camera. Photographic recording or the use of a photocell is, of course, essential for the detection of ultraviolet radiation. Similarly, some form of infrared detector is required for the infrared region. This can be a photocell; a thermocouple or thermopile; a bolometer, in which the thermal radiation causes changes in electrical resistance; or a pneumatic detector. In the Golay pneumatic detector a small gas-filled container undergoes a pressure change due to the heating effect of the radiation. One wall of the container is movable and is also reflecting. Thus the pressure changes produce a movement of this wall, causing a reflected beam of light to be deflected onto a photocell and so detecting changes in the incident infrared radiation.

Before using a spectrometer, it is first necessary to adjust the collimator to give a parallel beam and the telescope to receive it – that is, the telescope must be focused for infinity. This is most easily done with visible light and the instrument will then be in correct adjustment for use with ultraviolet or infrared radiation. Alternatively, if it is required to adjust the spectrometer directly for ultraviolet radiation using photographic recording, the eyepiece is removed and replaced by a camera focused for infinity; or a photographic plate may be mounted in the plane of the graticule. Similarly, any of the other detectors suitable for either ultraviolet or infrared use may be mounted at the graticule position. However, to adjust the spectrometer using visible light, it is first necessary to adjust the telescope eyepiece to the position where the graticule is sharply in focus when the eye is relaxed and focused at infinity. The rays are then made parallel by *Schuster's method*. To do this, the complete spectrometer is set up, with the prism on the spectrometer table and the prism and telescope rotated to the position for minimum deviation. The source of light for this setting-up procedure should be one that gives a distinctive isolated spectral line. The telescope is then rotated so as to increase the angle made with the 'straight-through' position of the collimator. The spectral line image is thus shifted to the edge of the field as seen through the telescope. This image is then brought back to the centre of the telescope field by rotating the prism. The prism may be rotated in either direction because there are two positions of the prism that give a central image [*Figure 7.2 (a)*]. When the prism is in position A, that is, the light is falling less obliquely onto the prism face, the slit-to-lens distance in the collimator is adjusted to obtain the sharpest image as seen through the telescope. The prism is then rotated to position B so that the rays from the collimator fall more obliquely onto the prism face and now the telescope is focused to obtain the sharpest image. This process is repeated several times – with position A adjust the collimator, with position B adjust the telescope – until the image appears equally sharp and without parallax on the graticule with the prism in either position. The collimator and telescope are then set for parallel rays.

It is also necessary to set the prism table level so that the refracting edge of the prism is parallel to the collimator slit and to the vertical rotation axis of the spectrometer. This may be done simply by arranging the prism as in *Figure 7.2 (b)*, where P, Q, and R are the three table levelling screws and A is the refracting angle of the prism. Face AC is made perpendicular to the line joining screws R and Q. Parallel light from the collimator is reflected from the faces AB and AC as shown. The telescope is first rotated to receive the reflected image of the collimator slit from face AC. Any of the levelling screws may be adjusted to

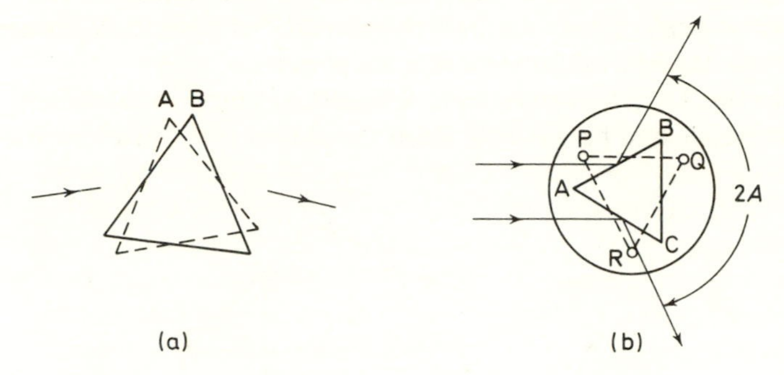

Figure 7.2. Illustrating the setting up of a prism spectrometer

centralise the image vertically in the field of the telescope. The telescope is then rotated to receive the reflection from face AB. Again, the image is centralised vertically but this time by the adjustment of the screw P only. This has the effect of tilting the reflecting face AB but only shifts face AC in its own plane and does not, therefore, upset the first setting. At this time the orientation of the collimator slit should also be adjusted so that it appears vertical with either reflection position.

Figure 7.2 (b) also shows that the angle between the reflected rays, as measured by the two angular positions of the telescope, is equal to twice the refracting angle A of the prism. Thus both the angle A and the angle of deviation for a given wavelength may be measured. The angle of minimum deviation may be determined either by measuring the angle between the minimum-deviation position of the telescope and the 'straight-through' position without the prism or, preferably, by rotating the prism and telescope to obtain the minimum-deviation position on the other side of the collimator axis and halving the angle between the two minimum-deviation positions of the telescope.

Using the refracting angle A and the angle of minimum deviation D, the refractive index n of the prism material for given wavelengths can be determined from

$$n = \frac{\sin \frac{1}{2}(A + D)}{\sin \frac{1}{2}A} \tag{7.6}$$

This is a particularly accurate method of determining the refractive index of a solid, providing it can be formed into a prism. By drawing a calibration graph of angles of deviation against known wavelengths, the arrangement may be used to determine unknown wavelengths. A particular series of wavelengths is characteristic of a particular element and thus may be used for identification purposes.

For larger spectrometers, there is insufficient mechanical stability if the telescope is to be rotatable about the central axis. Therefore, use is

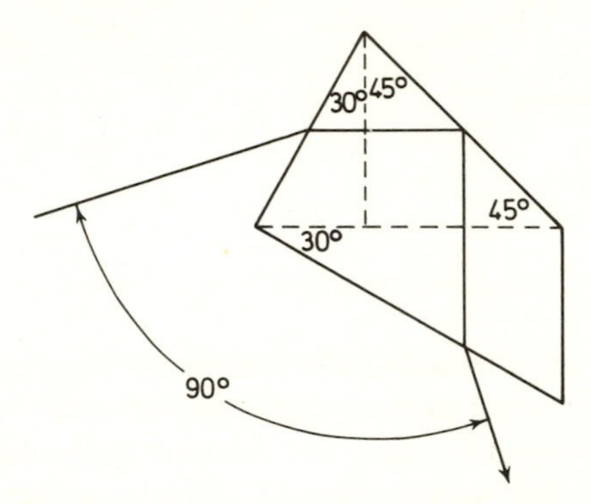

Figure 7.3. A constant-deviation prism

made of constant-deviation prisms. For example, the prism shown in *Figure 7.3* gives a minimum deviation of 90° for a refracted ray. With this arrangement, it is only necessary to rotate the prism to find the minimum deviation and it is thus the orientation of the prism that is measured rather than that of the telescope.

When using a diffraction grating or a prism to determine the wavelengths of spectral lines, it is important that adequate resolution is available. The resolving power of a grating or prism is its ability to distinguish two closely adjacent spectral lines as separate lines. If two lines have wavelengths λ and $\lambda + \delta\lambda$ and an angular separation $\delta\theta$, then by the Rayleigh criterion they are just resolvable when the diffraction maximum due to one wavelength coincides with the first diffraction minimum due to the other wavelength – that is, if the spectral line due to the wavelength λ and in a direction θ coincides with a minimum occurring immediately adjacent to a spectral line of wavelength $\lambda + \delta\lambda$ and of the same order m, as in *Figure 7.4 (a)*.

The angular separation $\delta\theta$ between the just-resolvable spectral lines is [*Figure 7.4 (a)*] the angular distance between a spectral line and its adjacent minimum. If θ is the direction of this spectral line and $\theta + \delta\theta$ is the direction of its adjacent minimum [*Figure 7.4 (b)*], the difference between the extreme path differences is λ. The path difference

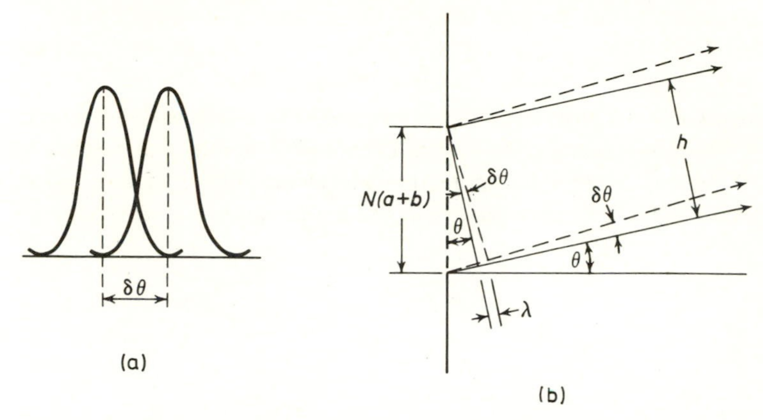

Figure 7.4. Diagram for calculating the resolving power of a diffraction grating

between rays from the top edge of the grating and from the middle is then $\lambda/2$ and so their interference effects cancel, as do the pair of corresponding rays just below these, and so on for all the pairs right across the grating. That is, a change in the extreme path difference of λ causes a change from a maximum of intensity to a minimum. Thus, from *Figure 7.4 (b)*,

$$\delta\theta = \frac{\lambda}{h} \tag{7.7}$$

for the spectral lines to be just resolvable, where h is the effective aperture of the grating. The difference in the wavelengths associated with this angle $\delta\theta$ is

$$\delta\lambda = \delta\theta \frac{d\lambda}{d\theta}$$

$$= \frac{\lambda}{h} \times \frac{(a+b)\cos\theta}{m} \tag{7.8}$$

by substituting from equations 6.11 and 7.7. Also

$$h = N(a + b) \cos \theta \qquad (7.9)$$

where $(a + b)$ is the grating space and N is the number of rulings. Therefore, substituting in equation 7.8

$$\delta\lambda = \frac{\lambda}{Nm}$$

i.e.

$$\frac{\lambda}{\delta\lambda} = Nm \qquad (7.10)$$

where m is the order of the spectrum and $\lambda/\delta\lambda$ is the resolving power.

Thus, for example, the sodium-D lines with wavelengths $5{\cdot}890 \times 10^{-7}$ m and $5{\cdot}896 \times 10^{-7}$ m require a resolving power, $\lambda/\delta\lambda$, of approximately 1000. To resolve these spectral lines requires a grating with a

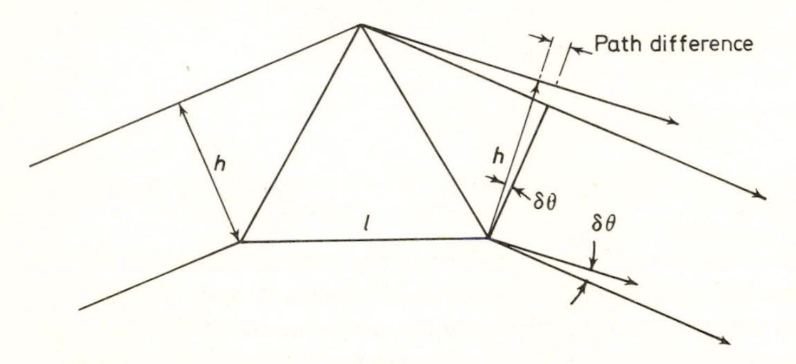

Figure 7.5. Diagram for calculating the resolving power of a prism

total of 1000 lines for the first order but only 500 for the second order. Very high resolving powers may be obtained using diffraction gratings; for example, a 50 mm wide grating with 1000 lines to the millimetre has a resolving power of 150000 in the third order.

The resolving power of a diffraction grating can be compared with that of a prism as follows. In *Figure 7.5* a plane wave front of width h is incident onto a prism of base length l such that the beam passes through at minimum deviation. If the incident wave front is composed of wavelengths λ and $\lambda + \delta\lambda$, then refraction by the prism produces two wave fronts, inclined to each other at an angle $\delta\theta$. If the prism material has refractive indices n and $n + \delta n$ for light of wavelengths λ and

$\lambda + \delta\lambda$ respectively, then the maximum optical path difference for rays through the prism is

$$l(n + \delta n) - ln = l\delta n \tag{7.11}$$

But from *Figure 7.5* this optical path difference is equal to $h\delta\theta$. Therefore

$$\delta\theta = \frac{l}{h}\,\delta n \tag{7.12}$$

The prism may also be regarded as a single diffraction slit of width h. Then, if $\delta\theta$ is the angular separation of two spectral lines of wavelength λ and $\lambda + \delta\lambda$, which are just resolvable – that is, $\delta\theta$ is the angle between the central maximum and the first minimum of the diffraction pattern – then by equation 7.7

$$\delta\theta = \frac{\lambda}{h}$$

and hence by equation 7.12

$$\lambda = l\delta n \tag{7.13}$$

Hence the resolving power is

$$\frac{\lambda}{\delta\lambda} = l\frac{dn}{d\lambda} \tag{7.14}$$

$dn/d\lambda$ is normally a negative quantity but must be treated here as positive since the resolving power is regarded as a positive number.

For ordinary flint glass $dn/d\lambda$ is of the order of 100 if λ is measured in millimetres – that is, the resolving power is approximately a hundred times the length of the prism base in millimetres. Thus, a prism with a base length of 10 mm can resolve the two sodium-D lines, providing that the spectrometer telescope has sufficient resolution and the other optical components are sufficiently free from aberrations.

The use of emission spectrometry in which the spectra are recorded photographically has the disadvantage that the specimen is destroyed and the measurement of intensity from the photographic plate can only be made to about ± 2% to ± 5% accuracy. If instead a spectrophotometer is used to measure absorption spectra, this accuracy can be increased to perhaps ± 0·2% using suitable photocells. Normally about 1×10^{-7} to 1×10^{-4} kg of sample is sufficient for absorption measurements. The technique has the advantage that the sample is generally recoverable and can be used for other tests.

A spectrophotometer comprises a source, which should be capable of providing radiation over the whole spectral region of interest; a monochromator to disperse this radiation and isolate a narrow wavelength region for transmission through the specimen; suitable mirrors to focus the transmitted radiation onto a photocell; and a suitable amplifier and chart recorder for recording the transmitted intensity with wavelength. A typical double-beam arrangement is shown in *Figure 7.6*.

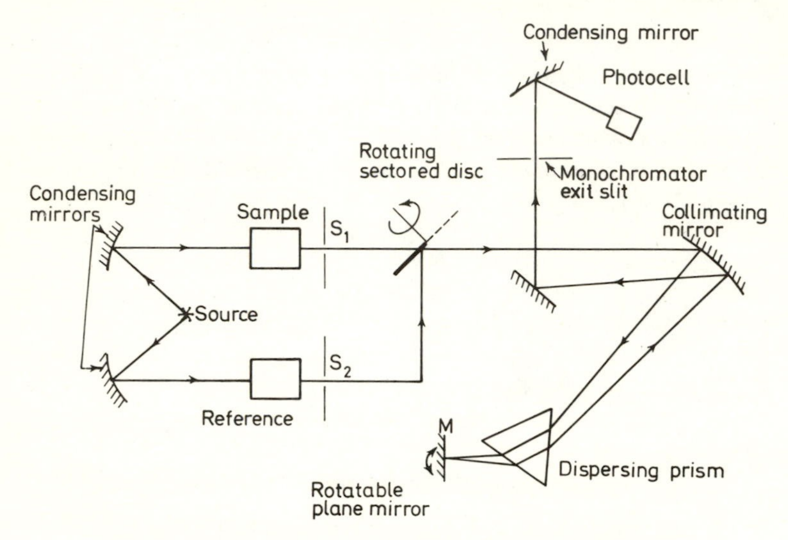

Figure 7.6. Arrangement of a photo-electric spectrophotometer

In this, light from the source is focused by two cylindrical mirrors onto slits S_1 and S_2. Close to the slits, where the two beams are narrow, are placed two cells, one of which contains the sample to be analysed with the other being a blank cell to act as a reference. The sample is generally in liquid form and may be dispersed in a suitable solvent, although solids and gases may also be studied. The two beams are then brought together in the plane of a rotating sectored disc, or chopper. This alternately allows the sample beam to be transmitted and the reference beam to be reflected towards a collimating mirror which reflects the beams so that they pass through a dispersing prism and onto the mirror M. This mirror reflects the two sets of absorption spectra so that part of the wavelength range passes through the monochromator exit slit. By rotating the mirror M, a different region of the spectrum is selected and, by reducing the slit width, a satisfactory wavelength resolution can be chosen, although this will also reduce the transmitted intensity. The transmitted narrow band of wavelengths is then detected by a photocell and any alternating signal amplified. If the sample and reference beams have the same intensity at a particular wavelength, then there will be no alternating signal produced by the chopper. If, however, the intensities are different owing to selective absorption by the sample, an alternating intensity beam will be transmitted. The resulting alternating signal received by the photocell will then be amplified and fed to a chart recorder which will record both the intensity due to the sample and the corresponding wavelength as determined by the angle of rotation of the mirror M.

A more accurate method is to use the amplified out-of-balance signal to move an attenuator across the reference beam until the out-of-balance signal is reduced to zero. The position of the attenuator is then a function of the intensity transmitted by the sample. By using the spectrophotometer as a null instrument in this way, errors due to variations in the radiation source and the photocell detector, as well as due to any drift in the gain of the amplifier, are eliminated. It is also important that the two beams have equal intensity before the sample is added. The attenuator thus also enables the beams to be initially balanced to cancel out variations between sample cells and, for example, the effects of solvents.

For ultraviolet and visible radiation use, incandescent lamps or discharge tubes are used as sources, whilst a Nernst filament or a Globar is normally used for an infrared source. A Nernst filament is a brittle high-resistance element made from sintered zirconium, cerium, and thorium oxides and a Globar is a rod of silicon carbide. These infrared sources are operated at temperatures up to 1800°C to give a satisfactory intensity of radiation. As has already been mentioned, the dispersing prism must be transparent to the radiation of interest, that is, calcium fluoride or lithium fluoride are suitable materials for the far-ultraviolet region, quartz for the near ultraviolet, rock salt for the infrared, or glass for the visible region.

7.4 X-ray Spectroscopy

As was discussed in Section 5.6, providing the excitation energy is sufficient, an element will be excited into emitting its characteristic X-ray spectrum. This spectrum consists predominantly of two lines, the Kα and the Kβ, although both are in fact multiplets. Also present are longer-wavelength lines, such as the L series. Even so, the characteristic X-ray spectrum of an element is considerably more simple than its optical counterpart. It is thus particularly suitable for the analysis of mixtures but has the disadvantage that generally more sample is required than with the optical methods. Optical spectroscopy is particularly suitable for the determination and measurement of very small amounts of impurity in an otherwise homogeneous sample but is less suited for the measurement of impurities present in relatively large proportions. In contrast, X-ray spectroscopy is generally of little use

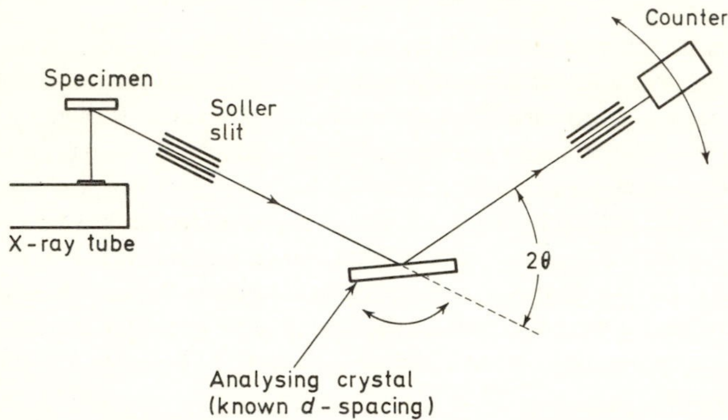

Figure 7.7. Arrangement of an X-ray spectrometer

with impurities present at less than about 5% of the total sample but is very suitable for the analysis of mixtures in which the components are present to similar extents.

In Section 6.6, reference has already been made to Bragg's law (equation 6.16)

$$2d \sin \theta = n\lambda \tag{7.15}$$

In accordance with this equation, a monochromatic X-ray beam of wavelength λ will be diffracted through an angle 2θ by a crystal lattice of spacing d, where $n = 1, 2, \ldots$. Thus, if the characteristic radiation of an element is diffracted by a crystal of known d-spacing, the wavelength of the radiation may be determined providing the diffracting angle can be measured. An advantage of X-ray spectroscopy is that a laboratory already equipped for X-ray diffraction with an automatic X-ray diffractometer of the type shown in *Figure 6.15* requires comparatively little additional equipment to convert it for spectroscopic use, as in *Figure 7.7.*

The X-ray tube is normally one with a tungsten target and operated to give a high-intensity continuous radiation. This radiation is then incident on the sample and is of sufficient intensity to excite its characteristic radiation. After being collimated by a Soller slit, this radiation is incident on an analysing crystal of known *d*-spacing. The crystal is one with a strong layer structure and is cut with its reflecting surface parallel to the layering. As with the diffractometer, the crystal is rotated in conjunction with a counter so that the counter is always in the correct position for a Bragg reflection, that is, whenever the condition of equation 7.15 is satisfied. As with the diffractometer, the output from the counter is electronically amplified and displayed on a chart recorder. Thus both intensities and the corresponding 2θ diffraction angles are readily measured and the wavelengths calculated from equation 7.15. From a table of characteristic wavelengths, the particular element may then be determined. For higher orders of n, the spectrum repeats itself so that care must be taken as a first-order spectral line may overlap a second-order line of a different element. As well as a chart recorder display, digital display facilities are usually available so that the spectrometer may be turned to a particular angle and an intensity count made on one line, rather than traversing the complete spectrum.

Although elemental analysis is simple with this technique, quantitative analysis of a mixture of components does present some problems. For example, one component of the mixture may be a strong absorber of the radiation from another of the components. Conversely, the characteristic radiation from one of the components may be such as to excite the characteristic radiation of another, so that this X-ray fluorescence, together with the primary excitation produced by the tungsten-anode X-ray tube, leads to enhancement of the second component. From a knowledge of the absorption coefficients and excitation energies required to excite fluorescence, corrections to the observed intensities may be made. In practice, however, it is generally more satisfactory to make up a series of known mixtures to bracket the unknown and hence interpolate to find the unknown proportions of the mixture. With some samples, such as a metal alloy, making a range of different proportional mixtures may present difficulties but often a powdered sample may be made up and compacted into a solid in a press, or sintered.

The specimen to be analysed is generally in solid form, preferably as a disc of about 20 mm diameter, but liquid samples may be analysed if held in a suitable cell. For example, with the arrangement shown in *Figure 7.7*, a liquid sample cell would have a thin plastic film base, such as of Mylar. It is important that the sample surface, in this case the bottom surface, be correctly located with reference to the incident X-ray beam and collimating Soller slit. The technique cannot be used with gas samples.

A disadvantage of X-ray spectroscopy is that long-wavelength X-rays are easily absorbed. Thus, for elements with atomic numbers below about 20 (calcium) in the Periodic Table it is necessary to evacuate the air from the apparatus. Even so, radiation from elements below 11 (sodium) in the Periodic Table are absorbed by the counter window. This means that elements such as carbon cannot be detected which is a serious disadvantage, especially for metallurgical use. Nevertheless, it must be regarded as a valuable supporting technique to X-ray diffraction as the spectrometer can determine the presence of most elements, assisting in the diffraction analysis which determines the distribution of these elements in compounds. It also has an advantage over optical spectroscopy in the speed with which results can be obtained with little specimen preparation. Thus, automatic X-ray spectrometers are used, for example, for quality control in steel works. The instrument may be programmed to move rapidly round to selected angles and count for particular elements; the analysis may then be given in the form of a print-out.

7.5 X-ray Microprobe Analysis

A combination of scanning electron microscope and X-ray spectroscopic techniques allows characteristic X-ray spectra to be obtained from very small samples. In the same way as an electron beam in a scanning electron microscope (Section 4.8) is focused down to a small spot and scanned back and forth to form a scattered electron image of the specimen, in a microprobe analyser a scanning electron beam is used to excite characteristic X-radiation which is used to produce an image in which one particular element shows up with high intensity.

The focused electron beam in a microprobe analyser has a spot diameter of about 1×10^{-6} m and, with the typical accelerating voltages of about 10–50 kV, the electron beam penetration into the specimen is of about the same order as the spot diameter. This means that the volume analysed is about 10^{-18} m^3. The electron beam excited characteristic X-radiation from this volume is passed through an analysing spectrometer and registered by a counter. Having set the spectrometer analysing crystal and the counter to the correct angles for the Bragg reflection of a particular wavelength, characteristic of a particular element, the X-ray signal received by the counter is used to modulate a cathode-ray tube in synchronisation with the scanning electron probe. Thus the distribution of a particular element over the surface of the specimen will show up as bright areas on the screen of the cathode-ray tube. The brightness gives a semiquantitative measure of the concentration of that element but, if an accurate measure is required, the scanning probe may be halted at the particular point of interest and the output from the counter fed into a ratemeter. Alternatively, the probe may be moved slowly along a line and the output from the ratemeter fed into a pen recorder. The scattered electrons may also be used to form a general image of the specimen, as in the scanning electron microscope. Thus a general image and an image showing the distribution of a given element may be produced side by side for comparison.

The specimen should normally be about 5 mm in diameter but the instrument can generally be adapted to take larger or smaller specimens. The specimen should preferably be electrically conducting to prevent build-up of electrostatic charge which would deflect and cause instability of the electron beam. The specimen chamber must also be evacuated owing to the requirements of the electron beam. Concentrations of less than 1 part in a 1000 are readily detectable so that the instrument is especially suitable for the analysis of precipitated phases in metals and alloys. Other uses for example occur in biology where inorganic inclusions such as silica in lungs and calcium in bones may be measured. Again, the absorption of long-wavelength X-rays sets a lower limit to the elements that may be detected. The limit with this technique is generally element 12 (magnesium), although some elements lower than this in the Periodic Table have been detected with specially constructed instruments.

7.6 *Other Spectroscopic Techniques*

Mention has already been made in Section 1.4 of mass spectrometers for the determination of partial pressures of gases. Since the instrument discriminates between different nuclear masses, it may also be used for analysis. A hot tungsten or rhenium filament is used as a source of electrons which are accelerated to collide with the gas molecules with sufficient energy to produce ionisation. The resulting positive gas ions are then accelerated by an applied potential V and constrained to move along a circular path by a magnetic field of intensity B. The force acting on the ion due to the magnetic field is then equal to Bev, where e is the charge on the ion and v its speed. By equating this force to the product of the mass and acceleration of the ion, the radius r of its trajectory is given by

$$Bev = \frac{mv^2}{r} \tag{7.16}$$

Equating the kinetic energy gained by the ion to its loss in potential energy, that is,

$$\tfrac{1}{2}mv^2 = eV \tag{7.17}$$

enables the radius r to be found from equations 7.16 and 7.17, that is,

$$r = \left(\frac{2mV}{eB^2}\right)^{1/2} \tag{7.18}$$

If the accelerating voltage V is varied, ions of different masses will have the correct trajectory to pass through the final slit and be recorded. Thus, by a suitable calibration of the voltage, the various ions may be identified. Monatomic gases such as argon may be ionised to form Ar^+, Ar^{2+}, . . ., while polyatomic molecules may be ionised to form various polyatomic ions. Thus, for example, a methane molecule, CH_4, may be ionised to form H^+, C^+, CH^+, $CH_2{}^+$, $CH_3{}^+$, and $CH_4{}^+$ ions. The method is especially suitable where small quantities of the substance to be analysed can be introduced as a gas into the evacuated mass spectrometer but some solids, such as the alkali metals, can be analysed by coating the filament with a salt containing the elements. By using an electric field in combination with a suitable magnetic field, a double-focusing arrangement may be achieved. With this, resolutions of the order of 1 part in 50000 may be obtained. Apart from the usual elemental analysis, a mass spectrometer may be used to gain knowledge of nuclear structure and binding energy from accurate mass determinations and the concept of the equivalence of mass and energy.

Another technique of spectroscopy is nuclear magnetic resonance (n.m.r.) which utilises the phenomenon exhibited by many atomic nuclei of distinct nuclear Zeeman energy levels being produced when in a magnetic field. This splitting of energy levels is based on the association of nuclear magnetic moments with quantised nuclear spins. The technique has been mainly limited to those nuclei of spin $\frac{1}{2}$. By applying a radio-frequency radiation, spectroscopic transitions can be induced between the energy levels. Thus the irradiating frequency may be varied in the presence of a constant magnetic field or, alternatively, the frequency may be maintained constant and the magnetic field varied. At a resonance condition, there will then be an absorption of the radio-frequency radiation. Although there will be the same frequency-to-field relationship between identical nuclei, the chemical arrangement of the nuclei within a sample will affect the net magnetic force and produce small differences in the transition frequencies. The particular resonant frequencies are thus a function of the molecular arrangement and enable individual molecules to be characterised.

A somewhat similar technique is that of electron spin resonance (e.s.r.), also termed electron paramagnetic resonance (e.p.r.), spectroscopy. This method is suitable for materials which show paramagnetism owing to the magnetic moment of unpaired electrons. At a resonance condition, more electrons will be moving up into a higher energy state than are moving down, so that an atom, when exposed to an oscillating magnetic field of fixed radio-frequency in the presence of a static magnetic field, will cause an absorption or dispersion of the energy of the oscillating field depending on the strength of the magnetic field. The technique is thus of use in the detection and identification of paramagnetic materials. It has also been used for the determination of electronic structure, molecular interaction studies, and for measurements of nuclear spins and moments. The method has been found to be of most use with solids, particularly solid solutions in single crystals, but has also found uses with liquids.

The technique of electron spectroscopy for chemical analysis (e.s.c.a.) involves the substance to be studied being bombarded with X-rays, ultraviolet radiation, or electrons. If the energy of the bombarding radiation exceeds the binding energy of electrons within a particular energy level of an atom or molecule, electrons will be ejected with a kinetic energy E_{kin} given by

$$E_{kin} = E_{rad} - E_b - E_r - W \tag{7.19}$$

where E_{rad} is the energy of the exciting radiation, E_b is the binding energy of the ejected electron, E_r the recoil energy, and W the work function involved. Both the last two terms may generally be regarded as negligible corrections. The ejected electrons then move along circular paths owing to an applied field and are focused on the slit of a detector. By varying the field strength, electrons of different kinetic energies may separately be recorded. At the correct energy conditions for the ejection of an electron, a resonance will occur leading to a marked increase in the number of electrons with a particular energy received at the detector. Only electrons from the outer 10^{-8} m thick layer of the specimen will have the energy expected from equation 7.19 as electrons from deeper in will be scattered away from the resonance condition. By suitably varying the energy of the exciting radiation, all the electron levels within the atom may be investigated, even in the presence of other atoms of a different species. Solids as well as gases may be studied by this technique and, since the binding energies will be slightly modified by the crystallographic arrangement of the atoms, information can be obtained concerning the distribution of charge within a molecule or, if this is known, concerning the structure of the molecule.

FURTHER READING

CHAPTER 1

Roberts, R. W., and Vanderslice, T. A., *Ultrahigh Vacuum and its Applications,* Prentice-Hall, New Jersey (1964)

Spinks, W. S., *Vacuum Technology,* Chapman & Hall, London (1963)

Yarwood, J., *High Vacuum Technique,* Chapman & Hall, London (1961)

CHAPTERS 2, 3, 4, AND 6

Bracey, R. J., *The Technique of Optical Instrument Design,* English Universities Press, London (1960)

Dyson, J., *Interferometry as a Measuring Tool,* Machinery Publishing Co., Brighton (1970)

Françon, M., *Optical Interferometry,* Academic Press, London (1966)

Marshall, S. L., *Laser Technology and Applications,* McGraw-Hill, New York (1968).

Tenquist, D. W., Whittle, R. M., and Yarwood, J., *University Optics* (2 vols), Iliffe, London (1970)

Tolansky, S., *Surface Microtopography,* Longmans, London (1960)

CHAPTERS 5 AND 6

Andrews, K. W., Dyson, D. J., and Keown, S. R., *Interpretation of Electron Diffraction Patterns,* Hilger & Watts, London (1967)

Cullity, B. D., *Elements of X-ray Diffraction,* Addison-Wesley, Reading, Mass. (1967)

Peiser, H. S., Rooksby, H. P., and Wilson, A. J. C., *X-ray Diffraction by Polycrystalline Materials,* Chapman & Hall, London (1960)

Rees, D. J., *Health Physics,* Butterworths, London (1967)

CHAPTER 7

Candler, C., *Practical Spectroscopy,* Hilger & Watts, London (1949)
Jenkins, R., and De Vries, J. L., *Practical X-ray Spectrometry,* Macmillan, London (1970)
Walker, S., and Straw, H., *Spectroscopy* (2 vols), Chapman & Hall, London (1966)

Appendix One

SI UNITS AND CONVERSION FACTORS

BASIC SI UNITS

Quantity	*Unit*	*Unit symbol*
Length	metre	m
Mass	kilogram	kg
Time	second	s
Electric current	ampere	A
Absolute temperature	kelvin	K
Luminous intensity	candela	cd

DERIVED UNITS HAVING SPECIAL NAMES

Quantity	*SI unit*	*Unit symbol*
Force	newton	$N = kg\ m\ s^{-2}$
Work, energy, heat	joule	$J = N\ m$
Power	watt	$W = J\ s^{-1}$
Electric charge	coulomb	$C = A\ s$
Electric potential	volt	$V = W\ A^{-1}$
Electric capacitance	farad	$F = A\ s\ V^{-1}$
Electric resistance	ohm	$\Omega = V\ A^{-1}$
Frequency	hertz	$Hz = s^{-1}$
Magnetic flux	weber	$Wb = V\ s$
Magnetic flux density	tesla	$T = Wb\ m^{-2}$
Inductance	henry	$H = V\ s\ A^{-1}$
Luminous flux	lumen	$lm = cd\ sr$
Illumination	lux	$lx = lm\ m^{-2}$

CONVERSION FACTORS

Length

$1\ \text{Å} = 10^{-10}$ m
1 in = 0·0254 m
1 ft = 0·3048 m
1 mile = 1·6093 km

Velocity

1 ft s^{-1} = 0·3048 m s^{-1}
1 mile h^{-1} = 0·4470 m s^{-1}

Area

1 in^2 = 6·4516 x 10^{-4} m^2
1 ft^2 = 0·0929 m^2
1 $mile^2$ = 2·5900 x 10^6 m^2

Volume

1 litre = 10^{-3} m^3
1 pint = 5·6826 x 10^{-4} m^3
1 UK gal = 4·5461 x 10^{-3} m^3
1 in^3 = 1·6387 x 10^{-5} m^3
1 ft^3 = 0·028 32 m^3

Mass

1 lb = 0·453 592 kg
1 ton = 1·01605 x 10^3 kg

Density

1 g cm^{-3} = 10^3 kg m^{-3}
1 lb in^{-3} = 2·7680 x 10^4 kg m^{-3}
1 lb ft^{-3} = 16·0185 kg m^{-3}

Force

1 dyn = 10^{-5} N
1 kgf = 9·8066 N
1 pdl = 0·1383 N
1 lbf = 4·4482 N
1 tonf = 9·9640 x 10^3 N

Pressure, stress

$$1\ \text{mmHg} = 133{\cdot}322\ \text{N m}^{-2}$$
$$1\ \text{torr} = 133{\cdot}322\ \text{N m}^{-2}$$
$$1\ \text{in H}_2\text{O} = 249{\cdot}1\ \text{N m}^{-2}$$
$$1\ \text{bar} = 10^5\ \text{N m}^{-2}$$
$$1\ \text{dyn cm}^{-2} = 10^{-1}\ \text{N m}^{-2}$$
$$1\ \text{kgf cm}^{-2} = 0{\cdot}0{\cdot}09807\ \text{N m}^{-2}$$
$$1\ \text{lbf in}^{-2} = 6{\cdot}8948 \times 10^3\ \text{N m}^{-2}$$
$$1\ \text{tonf in}^{-2} = 1{\cdot}5444 \times 10^7\ \text{N m}^{-2}$$

Viscosity

$$1\ \text{poise}\ (\text{g cm}^{-1}\ \text{s}^{-1}) = 10^{-1}\ \text{kg m}^{-1}\ \text{s}^{-1} = 10^{-1}\ \text{N s m}^{-2}$$
$$1\ \text{stokes}\ (\text{cm}^2\ \text{s}^{-1}) = 10^{-4}\ \text{m}^2\ \text{s}^{-1}$$

Magnetism

$$1\ \text{gauss} = 10^{-4}\ \text{T}$$
$$1\ \text{maxwell} = 10^{-8}\ \text{Wb}$$
$$1\ \text{oersted} = \frac{1}{4\pi} \times 10^4\ \text{A m}^{-1}$$

Radioactivity

$$1\ \text{curie (Ci)} = 3{\cdot}7 \times 10^{10}\ \text{s}^{-1}$$

Appendix Two

USEFUL PHYSICAL CONSTANTS AND VALUES

Quantity	*Symbol*	*Value*
Atomic mass unit		$1{\cdot}6604 \times 10^{-27}$ kg
Avogadro's constant	N_A	$6{\cdot}0225 \times 10^{26}$ kg-mole^{-1}
Boltzmann's constant	k	$1{\cdot}3805 \times 10^{-23}$ J K^{-1}
Electron charge	e	$1{\cdot}6021 \times 10^{-19}$ C
Electron rest mass	m_e	$9{\cdot}1091 \times 10^{-31}$ kg
Electron volt	eV	$1{\cdot}6021 \times 10^{-19}$ J
Gravitational constant	G	$6{\cdot}670 \times 10^{-11}$ N m^2 kg^{-2}
Planck's constant	h	$6{\cdot}6256 \times 10^{-34}$ J s
Standard gravitational intensity	g	$9{\cdot}8067$ m s^{-2}
Universal gas constant	R	$8{\cdot}3143 \times 10^3$ J kg-mole^{-1} K^{-1}
Velocity of light *in vacuo*	c	$2{\cdot}9979 \times 10^8$ m s^{-1}

Density of mercury at 0°C	$1{\cdot}3595 \times 10^4$ kg m^{-3}
Standard atmospheric pressure	$1{\cdot}0132 \times 10^5$ N m^{-2}
Vapour pressure at 20°C, mercury	$0{\cdot}1601$ N m^{-2}
Vapour pressure at 20°C, water	$2{\cdot}335 \times 10^3$ N m^{-2}
Velocity of sound at 20°C, air	$3{\cdot}434 \times 10^2$ m s^{-1}
Viscosity at 20°C, air	$1{\cdot}81 \times 10^{-5}$ N s m^{-2}
Viscosity at 20°C, water	$1{\cdot}0019 \times 10^{-3}$ N s m^{-2}
Young's modulus E, copper	$1{\cdot}298 \times 10^{11}$ N m^{-2}
Young's modulus E, steel	$2{\cdot}1 \times 10^{11}$ N m^{-2}

INDEX